Hydrogeology of the Mammoth Spring Groundwater Basin and Vicinity, Markagunt Plateau, Garfield, Iron, and Kane Counties, Utah

By Lawrence E. Spangler

Prepared in cooperation with the U.S. Forest Service

Scientific Investigations Report 2012–5199

U.S. Department of the Interior
U.S. Geological Survey

U.S. Department of the Interior
KEN SALAZAR, Secretary

U.S. Geological Survey
Marcia K. McNutt, Director

U.S. Geological Survey, Reston, Virginia: 2012

For more information on the USGS—the Federal source for science about the Earth, its natural and living resources, natural hazards, and the environment, visit http://www.usgs.gov or call 1–888–ASK–USGS.

For an overview of USGS information products, including maps, imagery, and publications, visit http://www.usgs.gov/pubprod

To order this and other USGS information products, visit http://store.usgs.gov

Suggested citation:
Spangler, L.E., 2012, Hydrogeology of the Mammoth Spring groundwater basin and vicinity, Markagunt Plateau, Garfield, Iron, and Kane Counties, Utah: U.S. Geological Survey Scientific Investigations Report 2012–5199, 56 p.

Contents

Figures

Plate

Tables

Conversion Factors, Datums, and Water-Quality Units

Multiply	By	To obtain
Length		
inch (in.)	2.54	centimeter (cm)
inch (in.)	25.4	millimeter (mm)
foot (ft)	0.3048	meter (m)
mile (mi)	1.609	kilometer (km)
Area		
square mile (mi^2)	2.590	square kilometer (km^2)
Flow rate		
cubic foot per second (ft^3/s)	0.02832	cubic meter per second (m^3/s)
gallon per minute (gal/min)	0.06309	liter per second (L/s)
Hydraulic conductivity		
foot per day (ft/d)	0.3048	meter per day (m/d)
Gradient		
foot per mile (ft/mi)	0.1894	meter per kilometer (m/km)

Temperature in degrees Celsius (°C) may be converted to degrees Fahrenheit (°F) as follows:

°F=(1.8×°C)+32

Temperature in degrees Fahrenheit (°F) may be converted to degrees Celsius (°C) as follows:

°C=(°F-32)/1.8

Vertical coordinate information is referenced to the National Geodetic Vertical Datum of 1929 (NGVD 29).

Horizontal coordinate information is referenced to the North American Datum of 1927 (NAD 27).

Altitude, as used in this report, refers to distance above the vertical datum.

Specific conductance is given in microsiemens per centimeter at 25 degrees Celsius (µS/cm at 25°C).

Concentrations of chemical constituents in water are given either in milligrams per liter (mg/L) or micrograms per liter (µg/L).

Stable-isotope (oxygen-18 and deuterium) concentrations are given in units of permil (per thousand). Radiochemical (tritium and gross alpha/beta) concentrations are given in units of picocuries per liter (pCi/L). Sulfur-35 concentrations are given in units of millibequerels per liter (mBq/L).

Wells by the Cadastral System of Land Subdivision

The well-numbering system used in Utah is based on the Cadastral system of land subdivision. The well-numbering system is familiar to most water users in Utah, and the well number shows the location of the well by quadrant, township, range, section, and position within the section. Well numbers for most of the State are derived from the Salt Lake Base Line and Meridian. Well numbers for wells located inside the area of the Uintah Base Line and Meridian are designated in the same manner as those based on the Salt Lake Base Line and Meridian, with the addition of the "U" preceding the parentheses.

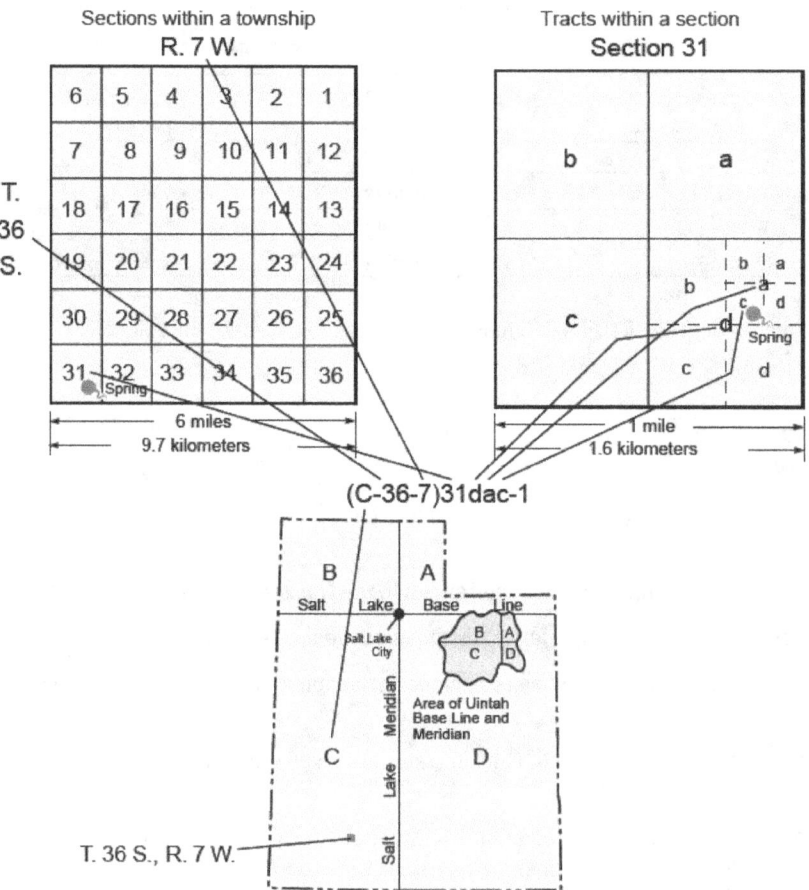

Surface-Water Sites— Downstream Order and Station Number

Since October 1, 1950, hydrologic-station records in U.S. Geological Survey reports have been listed in order of downstream direction along the main stream. All stations on a tributary entering upstream from a main-stream station are listed before that station. A station on a tributary entering between two main-stream stations is listed between those stations.

As an added means of identification, each hydrologic station and partial-record station has been assigned a station number. These station numbers are in the same downstream order used in this report. In assigning a station number, no distinction is made between partial-record stations and other stations; therefore, the station number for a partial-record station indicates downstream-order position in a list composed of both types of stations. Gaps are consecutive. The complete 8-digit (or 10-digit) number for each station such as 09004100, which appears just to the left of the station name, includes a 2-digit part number "09" plus the 6-digit (or 8-digit) downstream order number "004100." In areas of high station density, an additional two digits may be added to the station identification number to yield a 10-digit number. The stations are numbered in downstream order as described above between stations of consecutive 8-digit numbers.

Hydrogeology of the Mammoth Spring groundwater basin and vicinity, Markagunt Plateau, Garfield, Iron, and Kane Counties, Utah

By Lawrence E. Spangler

Abstract

The Markagunt Plateau, in southwestern Utah, lies at an altitude of about 9,500 feet, largely within Dixie National Forest. The plateau is capped primarily by Tertiary- and Quaternary-age volcanic rocks that overlie Paleocene- to Eocene-age limestone of the Claron Formation, which forms escarpments on the west and south sides of the plateau. In the southwestern part of the plateau, an extensive area of sinkholes has formed that resulted primarily from dissolution of the underlying limestone and subsequent subsidence and (or) collapse of the basalt, producing sinkholes as large as 1,000 feet across and 100 feet deep. Karst development in the Claron Formation likely has been enhanced by high infiltration rates through the basalt.

Numerous large springs discharge from the volcanic rocks and underlying limestone on the Markagunt Plateau, including Mammoth Spring, one of the largest in Utah, with discharge that ranges from less than 5 to more than 300 cubic feet per second (ft^3/s). In 2007, daily mean peak discharge of Mammoth Spring was bimodal, reaching 54 and 56 ft^3/s, while daily mean peak discharge of the spring in 2008 and in 2009 was 199 ft^3/s and 224 ft^3/s, respectively. In both years, the rise from baseflow, about 6 ft^3/s, to peak flow occurred over a 4- to 5-week period. Discharge from Mammoth Spring accounted for about 54 percent of the total peak streamflow in Mammoth Creek in 2007 and 2008, and about 46 percent in 2009, and accounted for most of the total streamflow during the remainder of the year.

Results of major-ion analyses for water samples collected from Mammoth and other springs on the plateau during 2006 to 2009 indicated calcium-bicarbonate type water, which contained dissolved-solids concentrations that ranged from 91 to 229 milligrams per liter. Concentrations of major ions, trace elements, and nutrients did not exceed primary or secondary drinking-water standards; however, total and fecal coliform bacteria were present in water from Mammoth and other springs. Temperature and specific conductance of water from Mammoth and other springs showed substantial variance and generally were inversely related to changes in discharge during snowmelt runoff and rainfall events. Over the 3-year study period, daily mean temperature and specific conductance of water from Mammoth Spring ranged from 3.4 degrees Celsius (°C) and 112 microsiemens per centimeter (µS/cm) during peak flow from snowmelt runoff to 5.3°C and 203 µS/cm during baseflow conditions. Increases in specific conductance of the spring water prior to an increase in discharge in 2008–09 were likely the result of drainage of increasingly older water from storage. Variations in these parameters in water from two rise pools upstream from Mammoth Spring were the largest observed in relation to discharge and indicate a likely hydraulic connection to Mammoth Creek. Variations in water quality, discharge, and turbidity indicate a high potential for transport of contaminants from surface sources to Mammoth and other large springs in a matter of days.

Results of dye-tracer tests indicated that recharge to Mammoth Spring largely originates from southwest of the spring and outside of the watershed for Mammoth Creek, particularly along the drainages of Midway and Long Valley Creeks, and in the Red Desert, Horse Pasture, and Hancock Peak areas, where karst development is greatest. A significant component of recharge to the spring takes place by both focused and diffuse infiltration through the basalt and into the underlying Claron limestone. Losing reaches along Mammoth Creek are also a source of rapid recharge to the spring. Maximum groundwater travel time to the spring during the snowmelt runoff period was about 7 days from sinking streams as far as 9 miles away and 1,900 feet higher, indicating a velocity of more than a mile per day. Response of the spring to rainfall events in the recharge area, however, indicated potential lag times of only about 1 to 2 days. Samples collected from Mammoth Spring during baseflow conditions and analyzed for tritium and sulfur-35 showed that groundwater in storage is relatively young, with apparent ages ranging from less than 1 year to possibly a few tens of years. Ratios of oxygen-18 and deuterium also showed that water from the spring represents a mixture of waters from different sources and altitudes. On the basis of evaluating results of dye-tracer tests and relations to adjacent basins, the recharge area for Mammoth Spring probably includes about 40 square miles within the Mammoth Creek watershed as well as at least 25 square miles outside and to the south of the watershed. Additional dye-tracer tests are needed to better define boundaries between the groundwater basins for Mammoth Spring and Duck Creek, Cascade, and Asay Springs.

Introduction

In October 2006, the U.S. Geological Survey (USGS) in cooperation with the U.S. Forest Service began a 4-year study to better understand the hydrology and water quality of Mammoth Spring on the Markagunt Plateau in southwestern Utah (fig. 1) and the relation between the contributing (recharge) area for the spring and adjacent springs, and the watershed in which the spring is located. Encroaching development along the margins of the watershed and increased recreational use and other activities within Dixie National Forest have raised concerns about potential effects on the spring, particularly with regard to water quality. Data from this study will provide land and water-resources managers, and others who utilize the watershed, the knowledge to recognize sensitive areas and potential effects, and to develop management alternatives for the protection of water quality, aquatic biota, and natural resources in the area.

Purpose and Scope

This report summarizes and interprets the results of an investigation of the hydrology of Mammoth Spring and the surrounding area and includes an analysis of discharge, water quality, and tracer data collected during the study. The report also describes the approach used to address the principal objectives of the study, which included (1) gaining a better understanding of the recharge area for Mammoth Spring and its relation to the surface-water drainage basin in which the spring discharges; (2) identifying potential point sources, such as losing streams and sinkholes where surface water can rapidly recharge and affect the aquifer that supplies the spring directly; (3) determining groundwater travel times through the aquifer and the relation between groundwater flow in the basalt and the underlying limestone; and (4) determining relations among precipitation, water quality, and discharge.

The scope of the study focuses on the hydrology of Mammoth Spring, but also includes data collected from other springs and surface-water sites on the plateau, especially in the Navajo Lake watershed, adjacent to and south of the Mammoth Spring watershed. Most of the data presented for Mammoth Spring were collected or obtained during the present study from November 2006 to December 2009; some additional data from 2010 and 2011 are also included. Prior to the current study, the USGS also made discharge measurements at the spring for the period 1954–57. Some water-quality data also have been collected intermittently by the USGS and are included here. In addition, water-quality data for Mammoth Spring that were collected by various state agencies and obtained from the U.S. Environmental Protection Agency (EPA) Storet database are included in the report. Although dye-tracer studies were carried out in the Navajo Lake watershed in the 1950s, which are summarized in this report, no previous tracer studies had been done in the Mammoth Spring watershed.

Previous Studies on the Markagunt Plateau

Wilson and Thomas (1964) investigated groundwater movement along the southern edge of the Markagunt Plateau, focusing on the hydrology of the Navajo Lake watershed, which lies immediately south of the Mammoth Spring watershed (fig. 1). Basalt flows have disrupted the natural surface-water courses in this area, resulting in subterranean piracy of the Navajo Lake outflow through sinkholes. Discharge increases at Cascade Spring and Duck Creek Spring in response to releases of water from Navajo Lake into the sinkholes indicated hydraulic connections between water lost from the lake and these springs, which were subsequently verified by dye-tracer tests during the investigation. Bifurcation of the groundwater flow path between the lake outflow and Cascade and Duck Creek Springs resulted in discharge to separate surface-water drainage basins. A more detailed discussion of the results of this investigation is presented in the "Dye-Tracer Studies" section of this report.

In 2002, the Mammoth Creek fish hatchery, located along the eastern margin of the Markagunt Plateau near the community of Hatch, became infected by whirling disease, caused by the microscopic parasite *Myxobolus cerebralis* (*http://whirlingdisease.montana.edu/*). To evaluate the potential for transport of whirling disease spores through a fractured basalt aquifer in the vicinity of the hatchery, the USGS, in cooperation with the Utah Division of Wildlife Resources, initiated a 3-year study in 2002 to determine hydrologic connections and groundwater travel times between losing reaches along Mammoth Creek, located about 10 mi downstream from Mammoth Spring, and springs at the hatchery (Spangler and others, 2005). On the basis of dye-tracer tests completed in October 2002 and October 2003, it was determined that water losing through the channel of Mammoth Creek about 3,000 ft southwest (upstream) of the hatchery discharged from the hatchery springs, with a groundwater travel time of about 7.5 hours.

Results of dye-tracer tests (Spangler and others, 2005) also indicated that groundwater travel time between Mammoth Creek and the hatchery springs is well within the 2-week timeframe of viability of the whirling disease parasite. Further, results of studies using cultured soil bacteria (*Acidovorax*) and club moss (*Lycopodium*) spores as surrogate tracers to simulate the size (about 30 microns) and movement of the parasite underground indicated that the potential exists for transport of the parasite from the creek to the springs. Although pathways of rapid groundwater flow were shown to exist between Mammoth Creek and the hatchery springs, low variability in springflow indicates that this is probably a small component of total discharge and that average groundwater travel time within the basalt aquifer is considerably longer.

Description of Study Area

The Markagunt Plateau, in southwestern Utah, lies at an altitude of about 9,500 ft within the Southern High Plateaus section of the Basin and Range-Colorado Plateau Transition province (Stokes, 1988) and covers an area of about 800 mi². The highest point on the plateau is Brian Head at 11,307 ft. The plateau is bounded on the west by the dramatic escarpment of Cedar Breaks National Monument, on the south by the Pink Cliffs, and on the east (not shown) by the Sevier River Valley (fig. 1). The principal surface drainage on the plateau is Mammoth Creek; however, much of the drainage from the plateau originates as springflow, which discharges to the Sevier River to the east, the Virgin River

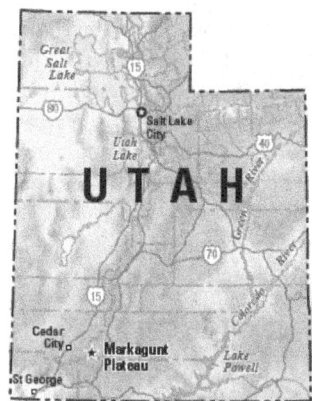

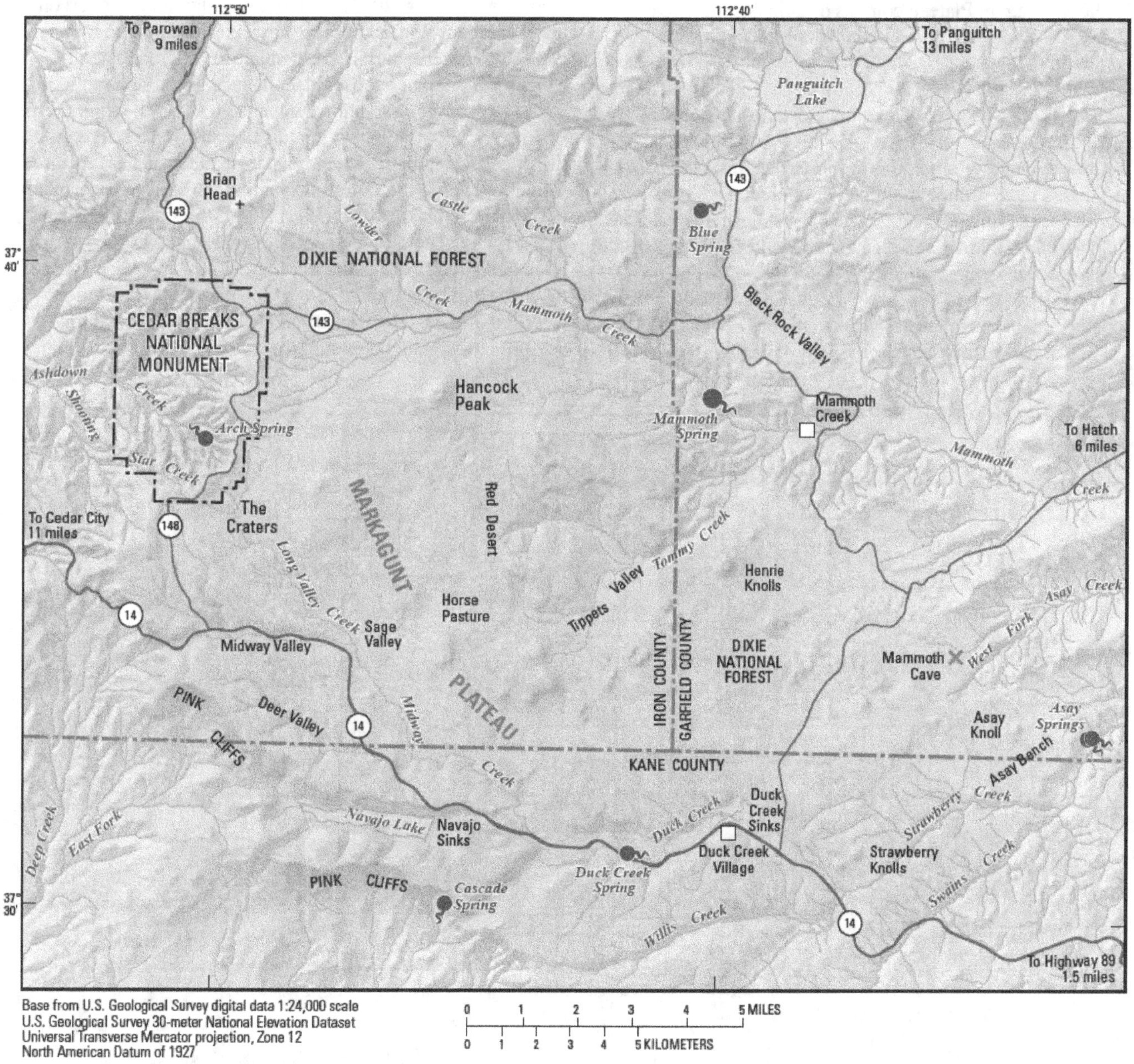

Figure 1. Location of study area and major physiographic features on the Markagunt Plateau in the vicinity of Mammoth Spring, southwestern Utah.

to the south, and Coal Creek in Cedar Canyon to the west. Navajo Lake, located along the southern margin of the plateau (fig. 1) at an altitude of about 9,000 ft, is unique in that it is almost entirely spring fed and all outflow from the lake is subterranean. Most of the surface of the plateau is included within Dixie National Forest. As a result, most of the population resides in small unincorporated communities such as Duck Creek Village and Mammoth Creek (fig. 1). Annual precipitation on the plateau averages about 30 in. (760 mm), mostly in the form of snow, which can reach a depth of 100 in. by early spring.

Geology

The Markagunt Plateau consists of a thick sequence of Cretaceous- and Tertiary-age sedimentary rocks that are overlain in many areas by Quaternary-age volcanic rocks, particularly basalt. The upper part of the sedimentary sequence consists of fine-grained calcareous sandstone, siltstone, mudstone, limestone, and minor conglomerate of the Paleocene- to Eocene-age Claron Formation (formerly the Wasatch Formation) that were deposited in fluvial, floodplain, and lacustrine (lake-deposited) environments. The Claron Formation is subdivided into a lower red member about 1,000 ft thick and an upper white member about 440 ft thick (Moore and others, 2004; Biek and others, 2011). In the Cedar Breaks area, exposed thickness of the Claron Formation is about 1,300 ft, of which about 1,100 ft consist of the red member (Gregory,

1950). Parts of the Claron Formation appear as massive, ledge- and cliff-forming beds of impure, locally cavernous limestone. These deposits make up the prominent escarpments on the south (Pink Cliffs) and west (Cedar Breaks National Monument) sides of the plateau (figs. 1 and 2) and dip gently to the east at about one and one-half degrees (140 ft/mi). They also compose the dramatic exposures of Bryce Canyon National Park, about 50 mi to the east on the Paunsagunt Plateau. Sinkholes developed in the Claron are numerous across the central part of the Markagunt Plateau, and the formation is capable of transmitting large amounts of water to springs, such as Cascade Spring along the Pink Cliffs (fig. 3).

Fluvial sandstone and mudstone of the Cretaceous- to Paleocene-age "formation of Cedar Canyon," an informal unit (Moore and others, 2004), underlie the Claron Formation on the Markagunt Plateau but are only exposed along the western and southwestern margins of the plateau, where erosion has cut deep enough to expose the unit, such as in the bottom of Cedar Breaks along Ashdown Creek and along the Pink Cliffs. Fluvial sandstone, siltstone, mudstone, and volcanic rocks of the Eocene- to Oligocene-age Brian Head Formation unconformably overlie the Claron Formation in the northern half of the plateau and are as much as 500 ft thick at Brian Head. These rocks generally are poorly exposed and weather to form large landslides that mantle the Claron, but have been stripped by erosion in the southern half of the plateau. At the northern edge of the study area, Brian Head strata are overlain by ash-flow tuffs, which are, in turn, overlain by the Markagunt

Figure 2. Outcrop of the red member of the Claron Formation in the Pink Cliffs along the southern margin of the Markagunt Plateau, southwestern Utah. View looking to the west along the Cascade Falls trail.

Figure 3. Discharge of Cascade Spring from the Claron Formation along the Pink Cliffs, Markagunt Plateau, southwestern Utah. The cave from which the spring discharges is developed along north, east, and northwest-trending joints and extends for more than 1,000 feet into the plateau.

megabreccia, a Miocene-age gravity-slide of regional extent (Biek and others, 2009, 2011) that consists of sedimentary and volcanic materials. Erosional debris derived from the mega-breccia locally blankets the upper reaches of the Navajo Lake and Mammoth Creek drainage basins.

Large parts of the Markagunt Plateau are capped by a veneer of Quaternary-age (mostly Pleistocene) volcanic rocks that directly overlie the Claron Formation. These rocks consist primarily of olivine basalts and andesites (Moore and others, 2004; Biek and others, 2007, 2009). The basalts are some of the youngest rocks in the state and form extensive sparsely vegetated lava flows throughout the region. Numerous cinder cones dot the surface of the plateau and are the sources for much of the lava. Some of the more prominent cones include Hancock Peak, Henrie Knolls, and Strawberry Knolls (figs. 1 and 4), which rise as much as 500 ft above the surrounding lava fields. Generally, thickness of individual flows is in the tens of feet; however, in areas where lava flows have filled valleys and other topographic lows, thickness may be several hundred feet. Volcanic activity probably began in the mid-Tertiary prior to and concurrent with regional uplift of the plateau, and the most recent eruptions (Quaternary age) occurred after

the current altitude of the plateau was attained (Wilson and Thomas, 1964).

Numerous normal faults have been mapped on the surface of the plateau (Moore and others, 2004; Biek and others, 2009). These generally trend north to northeast and have been traced for several miles in some areas. Displacement along the faults is generally small (tens of feet). This faulting is probably Pleistocene in age but does not cut through the younger basalt flows (Robert Biek, Utah Geological Survey, written commun., 2011). However, their surface expression is locally evident in the basalt by linear trenches or sinkhole alignments that result from dissolution along these faults in the underlying Claron Formation.

Karst Development

The land surface in some areas of the Markagunt Plateau can be characterized as a vulcano (pseudo) karstic terrain (Field, 2002). In these terrains, karst-like features can develop that are similar to those developed in limestone terrains, such as sinkholes and caves; however, these features typically are produced by non-solutional processes, such as surface collapses into lava tubes, which previously served as conduits for molten lava. In the southwestern part of the plateau, particularly between the Red Desert and an area known as The Craters (fig. 1), a unique terrain is present that is characterized by large sinkholes or dolines as much as 1,000 ft across and 100 ft deep (fig. 5). Most of these sinkholes are related to dissolution of limestone in the underlying Claron Formation and subsequent collapse and (or) subsidence of the basalt, rather than collapse into lava tubes. No outcrops of the Claron Formation have been observed in the bottoms of the deeper sinkholes, nor have shallow lava tubes been exposed in the walls of the sinkholes, implying that depth to the top of the Claron is greater than 100 ft or, more likely, that collapse of the basalt has obscured any exposures of the Claron. Some of these collapse features are distinctly elongate and appear to be aligned along fractures or faults developed in the underlying limestone along which dissolution has taken place (Moore and others, 2004). Sinkholes also are developed in the Claron Formation where it is not covered by basalt, particularly in the areas north and southeast of Navajo Lake. Sinkholes developed in the Claron, however, tend to be considerably smaller and shallower than those developed in areas where the formation has been covered by basalt. Karst development in the Claron probably began after the current altitude of the plateau was attained, when high precipitation and relief were present (Wilson and Thomas, 1964). Since extrusion of lava flows over the surface of the plateau during the Quaternary, however, runoff from the land surface has been substantially reduced by high infiltration rates through the basalt, and dissolution of the underlying limestone likely has been enhanced.

Sinking and losing streams are also typical of karst landscapes and are present in many areas on the plateau, although they generally are not obvious on topographic maps. Many of these streams are ephemeral, flowing only during the snowmelt runoff period, and their channels are dry during the

Figure 4. Hancock Peak cinder cone and lava flow, Markagunt Plateau, southwestern Utah. Note large sinkhole in basalt (left center). View is to the north.

Figure 5. Sinkhole in the Red Desert area of the Markagunt Plateau, southwestern Utah. Dissolution of the underlying Claron Formation has resulted in subsidence and (or) collapse of the basalt to depths up to 100 feet.

Figure 6.   Midway Creek losing all flow into channel deposits overlying the Claron Formation in the southwestern part of the Markagunt Plateau, southwestern Utah.

remainder of the year. Mammoth Creek and Tommy Creek lose water through unconsolidated channel deposits, particularly volcanic materials, most noticeably during the fall and winter months when snowmelt runoff no longer occurs in the entire channel and streamflow recedes upstream. Although these surface-water losses occur through stream channel deposits, in most cases, recharge is to the underlying Claron Formation, which generally lies within a few tens of feet, or less, of the land surface in these areas. Midway Creek, Long Valley Creek, and Duck Creek (fig. 1) terminate in "swallow holes" within their streambeds, where the entire flow of the stream is channeled underground into fractures or other voids in the underlying limestone. Streamflows as high as 19 ft³/s were measured terminating in swallow holes in the channel of Midway Creek during snowmelt runoff (fig. 6), and estimated streamflows of 4 to 5 ft³/s were observed terminating in swallow holes in the channel of Long Valley Creek. Wilson and Thomas (1964) reported flows of as much as 226 ft³/s at Duck Creek Sinks (pl. 1, sites 40, 41, 45). During the peak of snowmelt runoff, these swallow holes can be filled to capacity, and water flows overland in surface-water courses that are otherwise dry most of the year. Observations made along Midway Creek during this study also indicated that in the morning hours, when temperature is cooler, all flow is lost into the swallow holes, but later in the day, as temperature and, thus, snowmelt increases, the swallow holes can become filled to capacity, and overland flow occurs. Ephemeral streams also can form from overland flow during the snowmelt runoff period, which can then flow into nearby sinkholes.

A number of large springs discharge from the basalt or the underlying limestone on the Markagunt Plateau, including Cascade Spring, Arch Spring (in Cedar Breaks National Monument), Duck Creek Spring, Asay Springs, Blue Spring, and Mammoth Spring (fig. 1). Most of the major springs discharge laterally under gravity flow. Duck Creek and Blue Springs discharge as rise pools, whereas Cascade (fig. 3) and Arch Springs discharge directly from caves. Many springs, both large and small, discharge from multiple outlets or vents. Mammoth Spring discharges from numerous vents, many of which are active only during the snowmelt runoff period. At base (low) flow, the spring generally discharges from a broad area along the base of a hill. At higher flows, water levels near the spring rise, and additional vents begin flowing, some of which are several feet above the stage of the spring at baseflow. Generally, discharge of the major springs is highly variable, and peak flows can be 10 to 30 times baseflows. Peak flow is usually during the snowmelt runoff period in late May or early June, and baseflow occurs during the winter months.

Mammoth Spring, the focus of this study, is one of the largest springs in Utah, and has a discharge that typically ranges from less than 10 to over 200 ft³/s (fig. 7). A maximum instantaneous discharge of 314 ft³/s was recorded on June 6, 1957, at the peak of snowmelt runoff (Mundorff, 1971, fig. 4). On the basis of average discharge, Mammoth would be classified as a large second magnitude spring (Meinzer, 1927). Cascade, Duck Creek, and Lower Asay Springs have reported peak flows of about 25 to 35 ft³/s (Mundorff, 1971) but can have baseflows less than 1 ft³/s. Mammoth and Asay Springs are the major contributors to flow in the Sevier River, which flows to the north along the east side of the Markagunt Plateau. In addition, numerous smaller springs are present across the plateau, often forming the headwaters of tributaries

Figure 7.   Mammoth Spring at high flow during the snowmelt runoff period on the Markagunt Plateau, southwestern Utah. Springflow diminishes to less than 10 cubic feet per second during periods of baseflow.

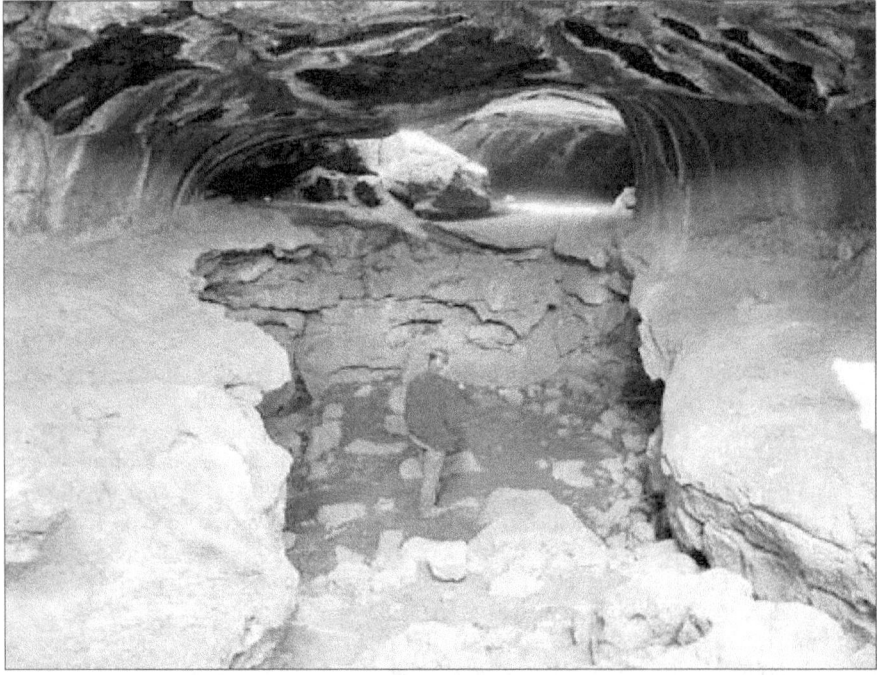

Figure 8.   Entrance passage in Mammoth Cave, a vulcanokarstic feature on the Markagunt Plateau, southwestern Utah. The roof of the lava tube has collapsed, allowing access to more than 2,000 feet of passage.

to the principal drainages. In some cases, the springs discharge from the toe of lava flows. These springs generally are intermittent or ephemeral but can have peak flows of several cubic feet per second during the snowmelt runoff period. Springs included in this category include Mammoth Creek rise pool (pl. 1, site 6) and Ephemeral spring (pl. 1, site 4), which discharge directly into Mammoth Creek upstream from Mammoth Spring. Although most of the major springs on the Markagunt Plateau discharge from the Claron Formation, the discharge points of the springs typically are mantled by volcanic and other colluvial materials.

Caves are present in both the basalt and the Claron Formation on the Markagunt Plateau. Duck Creek Lava Tube, one of the longest (over 12,000 ft) and highest altitude (about 8,560 ft) lava tubes in the continental United States, is located in the south-central part of the plateau and carries a small stream year-round. Other significant lava tubes on the plateau include Mammoth Cave, with about 2,200 ft of passage, and Bowers Cave, with almost 1,000 ft of passage, both of which are located in the south-central part of the plateau (pl. 1, sites 27 and 28). These vulcanokarstic features lie at very shallow depths, generally within 30 ft of the land surface, and access is typically through collapses in the roof of the lava tube (fig. 8). Most lava tubes in this area are associated with the most recent volcanic activity on the plateau and, thus, are very young features geologically. Although most do not contain flowing water, standing pools of water are not uncommon, particularly during and after the snowmelt runoff period, and perennial ice can be present in some of the lava tubes. Caves developed in the Claron Formation include those at Cascade Spring, located along the southern margin of the plateau in the Pink Cliffs, and Arch Spring, located along the western margin of the plateau in Cedar Breaks (fig. 1). These caves are some of the longest in the state in this formation, with each containing more than 1,000 ft of passage developed along prominent joints that have been enlarged by dissolution (fig. 3).

Methods

This investigation was carried out by using a multifaceted approach that included (1) field reconnaissance and inventory; (2) continuous stage (an indirect measurement of discharge) monitoring; (3) specific conductance (a surrogate for dissolved-solids concentration) and temperature monitoring; (4) water-quality sampling for major ions, trace elements, nutrients, and isotopes; (5) discharge measurements of springs; and (6) dye-tracer tests to determine subsurface connections and groundwater travel times.

Field Reconnaissance

A substantial amount of time was spent in field reconnaissance to locate significant recharge and discharge features, many of which are not identified on 7.5-minute scale topographic maps of the area. These features include direct surface-water inputs or focused points of recharge to the aquifer, such as losing and sinking (swallow holes) streams, and sinkholes that are termination points for losing streams (stream sinks). These features are potential entry points of contaminants into the aquifer and represent the source points, or origins, of some of the fastest flow paths within the aquifer and, thus, would be areas of greatest concern with respect to effects from anthropogenic activities. Numerous sinkholes also are present in the study area, particularly in the southwestern and south-central parts of the plateau, some of which were inventoried during this study. Many of the larger sinkholes are represented on the 7.5-minute scale Henrie Knolls and Navajo Lake quadrangle maps. Biek and others (2009, 2011) mapped sinkholes in this area by using aerial imagery and noted many not previously shown on topographic maps. Most of the sinkholes appear to have localized drainage areas that can be important capture areas for snowfall and subsequent recharge during the spring runoff.

Significant springs, which were defined for this study as those that are perennial or have discharges exceeding about 100 gal/min, were inventoried if encountered during reconnaissance. Springs not identified or labeled on topographic maps were assigned informal descriptive names and formatted in lower case (spring) for the purpose of this study, except in the case of spring nomenclature from previously published reports, which was retained. Locating significant springs was necessary for monitoring discharge points during dye-tracer tests, for determining hydrologic relations between adjacent springs, and for determining water-quality and discharge characteristics. Location and use data for 60 selected recharge and discharge sites that were inventoried and (or) monitored during the study are presented in table 1 and shown on plate 1.

Stage Monitoring and Discharge Measurements

Stage (relative water level) was recorded at Mammoth Spring on 1- and 2-hour intervals from November 2006 to December 2009. Continuous measurements were made to determine discharge variability of the spring, which was then used to help determine response to rainfall and snowmelt events and potential sources of water to the spring. Stage was recorded by using an In-Situ Inc., Troll 9000 series pressure transducer in conjunction with an In-Situ barotroll that was located near the spring to record barometric pressure. The Troll was installed approximately 100 ft downstream from the main springhead and below its high-water outlets, along the left bank (looking upstream) and above the confluence with Mammoth Creek. Because the pressure transducer was not vented, stage values were adjusted by using barometric pressure. Recorded stage was referenced to the height above the water surface of the top of an anchor rod (rebar) to which the Troll was mounted. This datum usually was reset in the software program (Win-Situ) for the transducer each time data were downloaded from the Troll, and a new monitoring cycle was begun. Stage data from the Troll were extracted onsite generally every 3 to 4 months.

Periodic measurements of the flow of Mammoth Spring were made by using a pygmy current meter in order to establish the relation between stage and actual discharge of the spring (appendix 1). These measurements were then used to establish a rating curve for determining intermediate values of discharge (Kennedy, 1983). Discharge measurements were made downstream from the Troll and upstream from the confluence with Mammoth Creek. To compare springflow and streamflow from the watershed, discharge of Mammoth Creek was measured just upstream from the confluence on the same day. During the snowmelt runoff period, when springflow was high and direct measurements could not be made, measurements were made downstream from the confluence at a location where the channel is wider, and the discharge of Mammoth Spring was obtained by subtracting the measured flow of Mammoth Creek above the confluence from the total measured flow. Periodic discharge measurements also were made at other springs and selected surface-water sites on the plateau to determine variability and relations between discharge and water-quality measurements (appendix 1).

Discharge of the combined flow of Mammoth Spring and Mammoth Creek is measured at USGS streamgaging station 10173450, "Mammoth Creek above west Hatch ditch, near Hatch, Utah," located approximately 8.5 mi downstream from Mammoth Spring. Discharge measurements at this gage for the period November 2006 to November 2007 were compared to the measured discharge of Mammoth Spring for the same period to evaluate runoff characteristics within the Mammoth Creek watershed and responses to snowmelt and rainfall events. Discharge measurements made at Mammoth Spring and Mammoth Creek above its confluence with the spring generally were made on the same day as streamflow measurements at the gage.

Table 1.   Location and use data for selected groundwater and surface-water sites on the Markagunt Plateau, southwestern Utah.

[deg, degrees; min, minutes; sec, seconds; NAD 27, North American Datum of 1927; NA, not applicable]

Map ID Refer to Plate 1	Site name	Township-Range-Section	Latitude (deg/min/sec) NAD 27	Longitude (deg/min/sec) NAD 27	County	Altitude (feet)	Topographic setting	Use of site	Surficial geologic unit	Permanence
1	Blue Spring	T 36S. R.7W. Sec. 18bda	37 40 58.5	112 40 35	Garfield	8,485	Large rise pool discharging from hillside	Significant discharge feature	Volcanic rock?	Perennial
2	Mammoth Creek at campground	T 36S. R.7W. Sec. 31acc	37 38 16.1	112 40 25.8	Garfield	8,170	Stream channel	Dye-injection site, discharge measurements	Channel deposits	Ephemeral
3	Mammoth Creek springs	T 36S. R.7W. Sec. 31bdb	37 38 28.7	112 40 48.1	Garfield	8,260	Discharges from multiple vents in bottom of streambed	Water-quality sample, water-quality measurements	Channel deposits	Perennial?
4	Ephemeral spring	T 36S. R.7W. Sec. 31bdb	37 38 29.1	112 40 48.9	Garfield	8,270	Discharges from shallow rise pit near Mammoth Creek	Water-quality sample, water-quality measurements, discharge, dye monitoring	Claron Formation?	Ephemeral
5	Mammoth Creek above Ephemeral spring	T 36S. R.7W. Sec. 31bdb	37 38 29.5	112 40 49	Garfield	8,265	Stream channel	Water-quality measurements	NA	Perennial
6	Mammoth Creek rise pool	T 36S. R.7W. Sec. 31bdd	37 38 22.8	112 40 34.8	Garfield	8,220	Discharges from rise pit near Mammoth Creek	Water-quality sample, water-quality measurements, discharge, dye monitoring	Claron Formation?	Ephemeral
7	Mammoth Creek above Mammoth Creek rise pool	T 36S. R.7W. Sec. 31bdd	37 38 23	112 40 35	Garfield	8,210	Stream channel	Water-quality measurements	NA	Ephemeral
8	Mammoth Creek below Mammoth Creek rise pool	T 36S. R.7W. Sec. 31bdd	37 38 22.5	112 40 34.5	Garfield	8,200	Stream channel	Water-quality and discharge measurements	NA	Perennial
9	Mammoth Spring at confluence with Mammoth Creek	T 36S. R.7W. Sec. 31dac	37 38 08	112 40 12	Garfield	8,120	Outflow from Mammoth Spring at junction with Mammoth Creek	Discharge measurements, dye monitoring	NA	Perennial
10	Mammoth Spring	T 36S. R.7W. Sec. 31dac	37 38 08	112 40 14	Garfield	8,125	Discharges from multiple vents along base of hillside	Water-quality sample, water-quality measurements, discharge, dye monitoring	Claron Formation	Perennial
11	Mammoth Creek at Mammoth Spring	T 36S. R.7W. Sec. 31dac	37 38 09	112 40 13	Garfield	8,120	Stream channel	Water-quality sample, water-quality measurements, discharge, dye monitoring	NA	Ephemeral
12	Mammoth Creek at upper injection site	T 36S. R.8W. Sec. 36abd	37 38 40.8	112 41 27.3	Iron	8,420	Stream channel	Water-quality sample, dye-injection site	Channel deposits	Ephemeral
13	Ashdown Creek below confluence	T 36S. R.9W. Sec. 29dcd	37 38 03.3	112 53 33.6	Iron	7,720	Stream channel draining Cedar Breaks amphitheatre	Discharge measurements, dye monitoring	NA	Perennial
14	Ashdown Creek above confluence	T 36S. R.9W. Sec. 29ddd	37 38 03.4	112 53 17	Iron	7,760	Stream channel draining Cedar Breaks amphitheatre	Dye monitoring	NA	Perennial
15	Shooting Star Creek	T 36S. R.9W. Sec. 29ddd	37 38 00.2	112 53 22.1	Iron	7,760	Stream channel draining Cedar Breaks amphitheatre	Dye monitoring	NA	Perennial
16	Arch Spring	T 36S. R.9W. Sec. 35dca	37 37 17	112 50 20.2	Iron	9,080	Discharges from cave near base of cliff in Cedar Breaks	Water-quality sample, water-quality measurements, dye monitoring	Claron Formation	Perennial
17	Asay Spring (lower)	T 37S. R.6W. Sec. 32daa	37 32 53	112 32 30.5	Garfield	7,120	Discharges from base of hillside along Asay Creek	Dye monitoring	Claron Formation	Perennial
18	Asay Spring (upper)	T 37S. R.6W. Sec. 32dac	37 32 51	112 32 36	Garfield	7,140	Discharges from base of hillside along Asay Creek	Dye monitoring	Claron Formation	Perennial
19	Mammoth Creek above highway	T 37S. R.7W. Sec. 4aca	37 37 35.5	112 38 8.5	Garfield	7,790	Mammoth Creek channel just upstream of Mammoth Creek highway	Dye monitoring	NA	Perennial
20	Mammoth Creek above Tommy Creek confluence	T 37S. R.7W. Sec. 4bbd	37 37 38	112 38 41.5	Garfield	7,840	Mammoth Creek channel above confluence with Tommy Creek	Dye monitoring and discharge measurements	NA	Perennial
21	Jensen springs	T 37S. R.7W. Sec. 4bcc	37 37 25.4	112 38 54	Garfield	7,900	Discharges from multiple vents in rise pool near Tommy Creek springs outflow	Water-quality and discharge measurements	Claron Formation?	Perennial

Table 1. Location and use data for selected groundwater and surface-water sites on the Markagunt Plateau, southwestern Utah.—Continued

[deg, degrees; min, minutes; sec, seconds; NAD 27, North American Datum of 1927; NA, not applicable]

Map ID Refer to Plate 1	Site name	Township-Range-Section	Latitude (deg/min/sec) NAD 27	Longitude (deg/min/sec) NAD 27	County	Altitude (feet)	Topographic setting	Use of site	Surficial geologic unit	Permanence
22	Mammoth Creek below Forest Service boundary	T 37S. R.7W. Sec. 5bbb	37 37 46	112 39 54.5	Garfield	8,040	Stream channel	Discharge measurements	NA	Perennial
23	Tommy Creek springs (weir)	T 37S. R.7W. Sec. 5dac	37 37 12.5	112 39 06.7	Garfield	7,935	Spring-fed tributary to Tommy Creek	Water-quality measurements	Claron Formation	Perennial
24	Tommy Creek springs (spring box)	T 37S. R.7W. Sec. 5dac	37 37 13	112 39 10.4	Garfield	7,940	Spring box alongside Tommy Creek	Water-quality measurements	Claron Formation	Perennial
25	Tommy Creek springs outflow	T 37S. R.7W. Sec. 5dac	37 37 16	112 39 05	Garfield	7,920	Combined flow from several springs in Tommy Creek drainage	Water-quality sample, water-quality measurements, discharge, dye monitoring	Claron Formation	Perennial
26	West Asay Creek spring	T 37S. R.7W. Sec. 25acd	37 33 56.8	112 34 52.3	Garfield	7,680	Discharges from side of valley along Asay Creek	Water-quality sample	Claron Formation	Perennial
27	Mammoth Cave	T 37S. R.7W. Sec. 25bdb	37 34 05	112 35 11.5	Garfield	7,925	Lava tube that is 2,200 feet long on flat area near West Asay Creek	Significant pseudokarst feature	Basalt	NA
28	Bowers Cave	T 37S. R.7W. Sec. 34bbd	37 33 19.6	112 37 38.8	Garfield	8,250	Lava tube that is 1,000 feet long in the Bowers Flat area	Significant pseudokarst feature	Basalt	NA
29	Big Spring	T 37S. R.8W. Sec. 1bda	37 37 42	112 41 39.5	Iron	9,080	Discharges at head of tributary drainage to Mammoth Creek	Significant discharge feature	Landslide deposits?	Perennial?
30	Tributary to upper Tommy Creek	T 37S. R.8W. Sec. 12add	37 36 33	112 40 59	Iron	8,620	Tributary to upper Tommy Creek drainage	Dye-injection site	Claron Formation	Ephemeral
31	Upper Tommy Creek springs	T 37S. R.8W. Sec. 13cba	37 35 34	112 41 49	Iron	8,840	Two springs discharging from base of hillside and toe of lava flow	Significant discharge feature	Basalt and Claron Formation	Ephemeral
32	Log cabin spring	T 37S. R.8W. Sec. 14aac	37 35 57.5	112 42 18	Iron	8,900	Discharges from hillside in upper reaches of Tommy Creek	Significant discharge feature	Claron Formation	Perennial?
33	Horse Pasture stream sink	T 37S. R.8W. Sec. 20aac	37 35 01.6	112 45 32.4	Iron	9,470	Shallow sink in Horse Pasture area	Significant recharge feature	Basalt	Ephemeral
34	Stream sink near old quarry	T.37S. R.8W. Sec. 30ccc	37 33 31	112 47 27.1	Iron	9,590	Shallow sink in Sage Valley area	Significant recharge feature	Claron Formation	Ephemeral
35	Long Valley Creek at swallow holes	T.37S. R.8 ½ W. Sec. 24cac	37 34 37.6	112 48 13	Iron	9,740	Multiple swallow holes in stream channel	Water-quality sample, dye-injection site	Channel deposits	Ephemeral
36	Midway Creek at swallow holes	T.37S. R.8 ½ W. Sec. 25bad	37 34 14.6	112 48 04	Iron	9,620	Multiple swallow holes in stream channel	Water-quality sample, discharge measurements, dye-injection site	Claron Formation	Ephemeral
37	Spring discharging from lava flow	T.37S. R.9W. Sec. 12abc	37 36 47.1	112 49 24.7	Iron	10,230	Discharges from toe of lava flow	Significant discharge feature	Basalt	Ephemeral
38	Stream sink along Highway 148	T.37S. R.9W. Sec. 14cca	37 35 22.4	112 50 55	Iron	10,180	Deep sink that is terminus for streamflow runoff along west side of Highway 148	Significant recharge feature	Claron Formation	Ephemeral
39	Stream sink near The Craters	T.37S. R.9W. Sec. 24abc	37 35 04	112 49 17.5	Iron	10,020	Sinkhole that is terminus for ephemeral stream	Dye-injection site	Claron Formation	Ephemeral
40	Duck Creek Sinks high water overflow	T 38S. R.7W. Sec. 5dcb	37 31 56	112 39 17.5	Kane	8,360	Collapse in basalt at north end of valley near Duck Creek Village	Significant source of recharge to Asay Spring	Basalt	Ephemeral
41	Duck Creek Sinks overflow	T 38S. R.7W. Sec. 5dcc	37 31 47	112 39 21	Kane	8,370	Sinkhole at base of hill at north end of valley in Duck Creek Village	Significant source of recharge to Asay Spring, dye-injection site	Basalt	Ephemeral
42	Duck Creek overflow into lava sink	T.38S. R.7W. Sec. 7aca	37 31 31.3	112 40 14.4	Kane	8,400	Overflow swallow hole into lava ridge 200 feet from Duck Creek	Significant recharge feature	Basalt	Ephemeral

Table 1. Location and use data for selected groundwater and surface-water sites on the Markagunt Plateau, southwestern Utah.—Continued

[deg, degrees; min, minutes; sec, seconds; NAD 27, North American Datum of 1927; NA, not applicable]

Map ID Refer to Plate 1	Site name	Township-Range-Section	Latitude (deg/min/sec) NAD 27	Longitude (deg/min/sec) NAD 27	County	Altitude (feet)	Topographic setting	Use of site	Surficial geologic unit	Permanence
43	Swallow hole on Duck Creek below Aspen Lake	T.38S. R.7W. Sec. 7aca	37 31 32.5	112 40 15	Kane	8,400	Swallow hole along south side of channel	Significant recharge feature	Claron Formation	Ephemeral
44	Duck Creek Lava Tube outflow	T.38S. R.7W. Sec. 7adb	37 31 28.1	112 40 08	Kane	8,400	Discharges from lava tube via culvert	Water-quality sample, water-quality and discharge measurements	Basalt	Perennial
45	Duck Creek Sinks	T.38S. R.7W. Sec. 8bac	37 31 33	112 39 31	Kane	8,370	Sinkhole at base of hill in Duck Creek Village	Significant source of recharge to Asay Spring	Claron Formation	Ephemeral
46	Navajo Lake rise pool 1	T.38S. R.8W. Sec. 7bad	37 31 35	112 47 00	Kane	9,035	Rise pool along north shore of Navajo Lake	Water-quality measurements, dye monitoring	Claron Formation	Perennial?
47	Navajo Lake rise pool 2	T.38S. R.8W. Sec. 8acb	37 31 29.5	112 45 52	Kane	9,035	Discharges from multiple vents in re-entrant along north shore of Navajo Lake	Water-quality measurements, dye monitoring	Claron Formation	Perennial?
48	Navajo Lake rise pool 3	T.38S. R.8W. Sec. 8acb	37 31 30	112 45 48	Kane	9,035	Discharges from three main depressions along north shore of Navajo Lake	Water-quality measurements	Claron Formation	Perennial?
49	Navajo Sinks	T.38S. R.8W. Sec. 8dab	37 31 17	112 45 33	Kane	9,020	Sinkholes developed in lakebed of Navajo Lake below dike	Significant source of recharge to Cascade and Duck Creek Springs, dye-injection site	Claron Formation	Ephemeral
50	Navajo lakebed springs	T.38S. R.8W. Sec. 9bac	37 31 36.1	112 45 00.2	Kane	9,090	A group of six springs discharging from hillside along northern shore of lakebed	Significant discharge feature	Claron Formation	Ephemeral
51	Site A21 spring	T.38S. R.8W. Sec. 12cbb	37 31 15.7	112 42 00.8	Kane	8,600	Ephemeral spring rising from lava tube in Forest Service campground	Significant discharge feature	Basalt	Ephemeral
52	Duck Creek Lake	T.38S. R.8W. Sec. 12cda	37 31 00.8	112 41 38.5	Kane	8,540	Sink point along northeast side of Duck Creek Lake	Dye-injection site	Basalt	Perennial
53	Duck Creek Spring	T.38S. R.8W. Sec. 12cdc	37 30 55.5	112 41 45	Kane	8,550	Discharges from large rise pool in Duck Creek Lake alongside Highway 14	Water-quality measurements, dye monitoring	Claron Formation	Perennial
54	Duck Creek Spring outflow (monitoring)	T.38S. R.8W. Sec. 12dcc	37 30 57	112 41 31	Kane	8,530	Outflow from Duck Creek Lake	Dye monitoring	NA	Perennial
55	Duck Creek Spring outflow (gaging)	T.38S. R.8W. Sec. 13aba	37 30 51.5	112 41 19	Kane	8,520	Outflow from Duck Creek Lake	Discharge measurements	NA	Perennial
56	Cascade Spring	T.38S. R.8W. Sec. 17dda	37 30 08	112 45 22	Kane	8,760	Discharges from cave near top of cliff along the Pink Cliffs	Water-quality and discharge measurements	Claron Formation	Ephemeral
57	Cascade Spring outfall	T.38S. R.8W. Sec. 17ddd	37 30 03	112 45 24.5	Kane	8,560	Instream pool 200 feet downslope from cave spring	Dye monitoring	Claron Formation	Perennial
58	Navajo Lake Spring	T.38S. R.8 ½ W. Sec. 1ccd	37 31 46	112 48 25.5	Kane	9,075	Discharges from hillside on west side of Navajo Lake	Water-quality and discharge measurements, dye monitoring	Claron Formation	Perennial
59	Deep Creek at Taylor Ranch	T.38S. R.9W. Sec. 7add	37 31 18.5	112 53 00.5	Kane	7,685	Stream channel along south side of Markagunt Plateau	Dye monitoring	NA	Perennial
60	Three Creeks at Larson Ranch	T.38S. R.9W. Sec. 18aad	37 30 42	112 52 56	Kane	7,650	Stream channel along south side of Markagunt Plateau	Dye monitoring	NA	Perennial

Water-Quality Monitoring

Water temperature and specific conductance were recorded at Mammoth Spring on 1- and 2-hour intervals from November 2006 to December 2009. Continuous measurements were made in conjunction with the stage measurements to determine the range in these parameters seasonally and to observe changes in temperature and specific conductance with variations in discharge. Data from the Troll 9000 series probe were extracted onsite generally every 3 to 4 months. Temperature and specific conductance readings from the Troll were checked against a calibrated handheld meter at the same time. A 100 µS/cm standard was used to check the Troll conductance sensor; the sensor was recalibrated when there was a discrepancy of more than 2 percent. Additional periodic measurements of temperature, specific conductance, and pH were made at the springhead, upstream from the Troll, and are presented in appendix 1. Miscellaneous measurements of temperature, specific conductance, and pH from other groundwater and surface-water sites on the plateau also are presented in appendix 1. In addition, continuous measurements of water temperature were made from April 2008 to November 2009 at a rise pool upstream from Mammoth Spring (pl. 1, site 6) by using Onset StowAway Tidbit temperature loggers. These measurements were used to evaluate relations between the spring and adjacent Mammoth Creek and to help determine periodicity of the spring.

Water-Quality Sampling

Water samples were collected from Mammoth and other springs and surface-water sites across the Markagunt Plateau and analyzed for major-ion chemistry, alkalinity, dissolved-solids concentration, selected trace elements, and nutrients, including nitrate plus nitrite, ammonia, and orthophosphate. Major-ion analyses were used to determine the general chemistry of water discharging from various locations on the plateau to help differentiate between water discharging from the Claron Formation and the overlying basalt, and to establish a baseline against which changes in water quality can be compared. Samples from Mammoth Spring were collected at low, moderate, and high flows to compare variations in chemistry with changes in discharge. Selected samples were collected from losing (sinking) streams to determine differences between source water chemistry and water discharging from Mammoth Spring.

Water samples were collected directly from spring sources and processed onsite according to procedures outlined in the USGS National Field Manual for the collection of water-quality data (U.S. Geological Survey, variously dated). Samples were pumped through 0.45-micron pore size capsule filters and collected in polyethylene bottles. Samples for analysis of cations and trace elements were stabilized with nitric acid to a pH of about 2. Total alkalinity, which is used to calculate bicarbonate and carbonate concentrations, was determined in

the field and in the laboratory by titration techniques. Samples collected for analysis of nutrients were chilled and sent to the laboratory within 48 hours of collection. Two quality-assurance inorganic blank water samples were processed onsite along with the environmental samples to evaluate equipment cleaning procedures. These samples were processed by using the same procedures as those used for the environmental samples and were analyzed for major ions. All water samples were analyzed by the National Water Quality Laboratory in Denver, Colorado, according to procedures outlined in Fishman and Friedman (1989). At the time of sample collection, temperature, pH, and specific conductance were measured at the location of sampling. Results of analysis of water samples collected during this study are stored in the USGS National Water Information System (NWIS) database (*http://ut.water.usgs.gov/infodata/waterquality.html*).

Water samples also were collected from Mammoth Spring and other selected springs for analysis of total and fecal coliform bacteria to evaluate contamination of the springs from surface-water sources and the potential for movement of water-borne diseases along groundwater flow paths. Samples were collected in 100-milliliter (mL) plastic bottles and immediately chilled to inhibit growth of bacteria. Samples were hand-delivered to the state of Utah (Unified) Health Laboratories in Salt Lake City within 24 hours of collection and were analyzed by using the most probable number (MPN) method (*http://bcn.boulder.co.us/basin/data/NEW/info/FColi.html*). Three samples collected from Mammoth Spring, Mammoth Creek rise pool, and the outflow from Duck Creek Lava Tube, also were collected and transported back to the Utah Water Science Center (UWSC) laboratory in Salt Lake City where they were analyzed by using the membrane filter (MF) method.

Water samples from selected sites were collected and analyzed for the stable isotopes of oxygen-18 and deuterium, and the radioisotopes of tritium (^{3}H) and sulfur-35, to help determine sources, mixtures, and ages of relatively longer-term components (1 to 50 years) in the aquifer. Samples for oxygen-18 and deuterium were collected from Mammoth Spring, as well as other selected springs and surface-water sources, to determine potential sources of water to the springs and hydraulic relations between Mammoth Creek and several springs along the creek. Water samples for oxygen-18 and deuterium were collected in 60-mL clear glass bottles and shipped to the USGS Stable Isotope Laboratory in Reston, Virginia, for analysis. A 1-liter (L) polyethylene bottle was used to collect a tritium sample from Mammoth Spring during baseflow conditions and analyzed at the USGS Isotope Laboratory in Menlo Park, California. Tritium concentrations peaked in the atmosphere during the 1960s and have since declined to levels that are considered background or "modern." Because the half-life of this isotope is about 12.5 years, non-detectable concentrations of tritium in groundwater indicate a component of water that is older than about 50 years. Samples for sulfur-35 were collected from Mammoth Spring at various stages of flow to determine components that have very short (less than 2 years) groundwater residence times. Sulfur-35 originates from precipitation that falls as rain or snow and enters the aquifer. Sulfur-35 is characterized by a half-life of about 87 days; thus, the presence of this isotope in groundwater indicates a component of flow that likely represents recharge to the aquifer during the previous year or snowmelt runoff cycle. Samples for sulfur-35 were collected in a 5-gallon container, stabilized to a pH of about 2, and processed through a resin cartridge to extract sulfate from the water, which was subsequently analyzed at the USGS Isotope Laboratory in Menlo Park, California. In addition, a sample was collected from Mammoth Spring during baseflow conditions and sent to Eberline Analytical Services in Richmond, California, for analysis of gross alpha/gross beta activity to determine levels of radioactivity in the groundwater.

Dye Tracing

Sodium fluorescein (uranine) and rhodamine WT water tracing dyes, and the optical brightener Tinopal CBS-X were used as groundwater tracers to establish hydrologic connections between surface-water inputs and Mammoth and other springs, help define groundwater basin boundaries between springs, and to determine groundwater travel times. These tracers were selected because of their relatively conservative nature in the environment, detectability at low concentrations and over long distances, ease of analytical detection, and very low toxicity. Fischer Chemical activated charcoal (6-14 mesh) contained in nylon-screen packets that were suspended on wires embedded in concrete weights, was used for adsorption of the fluorescein and rhodamine WT dyes. Undyed cotton linen mounted on embroidery hoops that were also suspended on the concrete weights, was used for adsorption of the optical brightener. These passive methods can be used to determine approximate (maximum) groundwater travel times and to establish connections between surface-water inputs, such as losing streams and sinkholes, and springs.

Charcoal and cotton detectors generally were collected and exchanged within the first 6 weeks after the initial dye injection and at longer intervals thereafter. Charcoal detectors were collected and exchanged year-round during the study period to track the residence time of the dye through the aquifer and to determine relative concentration levels in the spring water prior to subsequent injections of the same dye. During collection, the detectors were placed into labeled ziplock baggies and brought back to the UWSC laboratory, where generally, they were kept refrigerated to minimize dye degradation until analyzed. Charcoal samples were removed from the screen wire packets, placed into 100-mL glass beakers, and thoroughly rinsed with de-ionized water to remove dirt and organic debris. Fluorescein dye was extracted from the activated charcoal by using a 5-percent solution of potassium hydroxide and 70 percent isopropyl alcohol, a common eluent for extraction of this dye (Alexander and Quinlan, 1992). The samples were analyzed by using visual methods; if needed, the samples were analyzed on a Turner Model 10

filter fluorometer located in the UWSC laboratory. Rhodamine WT dye was extracted from the activated charcoal by using a mixture of 50 percent 1-propanol, 20 percent ammonium hydroxide, and 30 percent de-ionized water (Alexander and Quinlan, 1992). Selected eluted charcoal samples containing very low concentrations of rhodamine WT were sent to the Edwards Aquifer Authority Laboratory in San Antonio, Texas, for confirmation on a Perkin-Elmer scanning spectrofluorometer. Cotton fabric detectors also were labeled, bagged, and brought back to the laboratory, where they were rinsed with tap water, then allowed to dry. Optical brightener was qualitatively assessed by exposure of the cotton fabric to a handheld long-wave ultraviolet lamp and observation of the characteristic bluish-white fluorescence. For purposes of this study, results for all tracer tests were reported qualitatively as either detected (positive) or not detected (negative).

The amount of dye used for each injection was determined by using a formula derived from more than 5,000 dye-tracer tests in the United States in conduit-dominated karst systems (Worthington, 2007), as well as from personal experience in other karst terrains. The principal factors for determining the required amount of dye included the discharge of Mammoth and other springs, the straight-line length of the flow path from input to output points plus an additional 30 percent to accommodate sinuosity of the flow path, the estimated rate of groundwater flow (velocity), and the desired peak concentration of the tracer at the likely point of discharge. All significant springs, in addition to Mammoth Spring, were monitored during dye-tracer tests to determine if bifurcations in groundwater flow existed.

Hydrogeology of the Mammoth Spring Groundwater Basin and Vicinity

Groundwater Chemistry

Measurements and results of analyses for water-quality parameters, major-ion chemistry, and selected trace elements were used to help characterize and determine similarities among groundwater and surface-water sites on the Markagunt Plateau and to determine relations with changes in discharge. Analyses for nutrients and bacteria were used to assess the potential for contamination of springs from surface-water sources. Stable and radioisotopes were used to determine groundwater age and sources of water to Mammoth Spring and other springs within the study area.

Water-Quality Parameters

Temperature, specific conductance, and pH measurements for sampled groundwater and surface-water sites on the plateau are shown in table 2, and additional measurements of these parameters for selected sites are included in appendix 1.

Water temperature of springs ranged from 2.9 to 9.4°C, but variability differed for individual springs. The greatest variability in spring water temperature, with the least variability in discharge, was noted in Mammoth Creek rise pool (site 6) and Ephemeral spring (site 4), two rise pools located 0.5 and 0.75 mi upstream from Mammoth Spring, respectively (pl. 1, inset). Measured and estimated discharge of these springs ranged from zero flow to as much as 3 ft³/s during the study period, while the range in temperature spanned as much as 8°C (appendix 1). Ephemeral spring generally flows only from early spring to mid-summer and was observed on several occasions to exhibit a diurnal-like flow regime, where discharge from the shallow rise pit only occurred later in the day when groundwater levels in the vicinity of the spring rose high enough to initiate flow.

Continuous (1-hour interval) temperature measurements, expressed as daily mean values from April 2008 to December 2009, are shown in figure 9 for Mammoth Creek rise pool. The peak temperature of water from the spring reached about 10°C in late summer of 2008 and 2009, which is anomalously high for groundwater at this altitude when compared to the temperature of water from other springs in the area and is about twice as high as that of Mammoth Spring. In both years, the plot shows an overall increase in temperature beginning with the onset of snowmelt runoff, which contrasts to the more typical decrease in temperature that was documented in Mammoth (refer to discussion under "Relation between

Water Quality and Discharge") and other large springs in the region. Other characteristics of the rise pool temperature plot include periods of no flow from the spring, periods when the water level appears to be fluctuating within, or moving in to and out of the rise pit (about 4 ft deep), periods of intermittent flow from the rise pit, periods when the sensor is recording air temperature because the water level is below the bottom of the pit, and even a period of constant temperature that is attributed to snow cover in the rise pit (fig. 9).

High temperature variations in spring waters, even in terrains where springs are under surface-water influence, are unusual, particularly where the discharge variability is very low, as documented in Mammoth Creek rise pool and Ephemeral spring. The temperature of water entering the aquifer through focused points of recharge such as sinkholes generally is dampened as groundwater moves through the aquifer and mixes with water that has equilibrated to the temperature of the surrounding rock. The substantial variability in temperature of water from these springs is likely related to input from Mammoth Creek, along which both springs are located (within 100 ft in both cases) and which can vary widely in temperature diurnally as well as seasonally (appendix 1). Additionally, observed changes in turbidity of water from Mammoth Creek rise pool on August 16–17, 2007, in response to similar changes in Mammoth Creek following a significant rainfall event, further indicate a likely hydraulic connection between the creek and this spring.

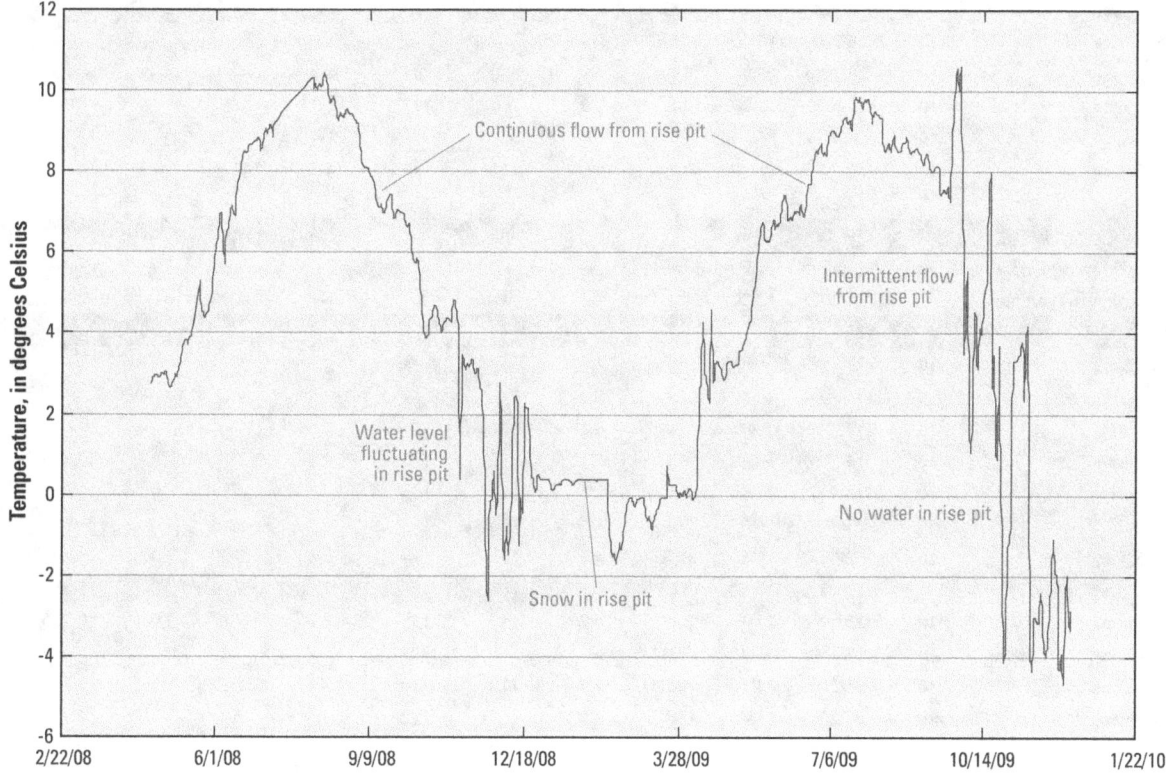

Figure 9. Water temperature and fluctuations in Mammoth Creek rise pool, Markagunt Plateau, southwestern Utah, April 2008 to December 2009.

Table 2.   Field measurements and results of chemical analyses for major ions, nutrients, and bacteria for selected groundwater and surface-water sites on the Markagunt Plateau, southwestern Utah.

[mm/dd/yyyy, month/day/year; hh:mm, hour minutes; ft³/s, cubic feet per second; °C, degrees Celsius; µS/cm, microsiemens per centimeter; NTU, nephelometric turbidity units; mg/L, milligrams per liter; ANC, acid neutralizing capacity; MPN, most probable number; ml, milliliters; E, estimated; —, not analyzed; <, less than; >, greater than]

Site name	Map ID Refer to Plate 1	Sample date (mm/dd/yyyy)	Time (hh:mm)	Discharge, ft³/s	Water temperature, °C	Specific conductance, field, µS/cm	Specific conductance, lab, µS/cm	pH, field, standard units	pH, lab, standard units	Turbidity, field, NTU	Turbidity, lab, NTU	Dissolved oxygen, field, mg/L
Mammoth Creek at Mammoth Spring	11	11/7/2006	16:10	1.49	3.7	236	238	8.5	7.9	—	—	—
Mammoth Creek rise pool	6	11/9/2006	13:30	1.13	3 9	239	242	7.8	7.8	—	—	—
		10/30/2008	15:00	E 0.28	4.4	242	—	7.7	—	—	—	—
Ephemeral spring	4	4/19/2007	21:30	0.40	2 9	224	237	7.9	7.8	—	—	—
Tommy Creek springs outflow	25	8/20/2007	20:15	1.45	5 1	168	178	8	8.1	—	—	—
Arch Spring	16	9/14/2007	16:10	<1	4.6	291	302	8.2	8	—	—	—
Mammoth Creek springs	3	7/10/2008	10:00	E 1–2	9.4	205	213	7.7	7.8	—	—	—
Midway Creek at swallow holes	36	5/20/2009	19:40	E 4	16 5	230	242	8.2	7.9	—	—	—
Long Valley Creek at swallow holes	35	6/1/2009	15:45	E 2	12.8	289	309	8.3	8.3	—	—	—
West Asay Creek spring	26	8/21/2009	10:30	E 1.25	6.8	364	382	7.6	7.8	—	—	—
Duck Creek Lava Tube outflow	44	8/21/2009	13:45	0.44	8 5	251	265	7.5	7.6	—	—	—
		10/30/2008	11:15	E 0.22	—	244	—	7.9	—	—	—	—
Mammoth Creek at upper injection site	12	11/12/2009	17:00	0.50	3 9	218	222	8.3	8.3	—	—	—
Mammoth Spring	10	7/14/1954	—	—	4 5	152	—	7.3	—	—	—	—
Mammoth Spring	10	8/6/1954	—	—	4 5	152	—	7.9	—	—	—	—
Mammoth Spring	10	9/28/1968	—	—	6	158	—	7.5	—	—	—	—
Mammoth Spring	10	6/17/1989	16:30	—	5	170	157	7.2	8.4	—	—	—
Mammoth Spring	10	11/7/2006	14:45	11.7	4 3	164	168	7.9	8.1	—	—	—
Mammoth Spring	10	2/20/2007	18:30	5.8	—	—	—	—	—	—	—	—
Mammoth Spring	10	3/26/2007	16:00	18.5	—	—	—	—	—	—	—	—
Mammoth Spring	10	2/19/2008	15:00	5.8	—	—	—	—	—	—	—	—
Mammoth Spring	10	5/2/2008	13:00	88	3 9	150	137	8.1	8.1	—	—	—
Mammoth Spring	10	10/30/2008	14:15	7.7	—	—	—	—	—	—	—	—
Mammoth Spring	10	5/21/2009	11:30	200	4	137	144	8.1	8	—	—	—
Mammoth Spring	10	4/28/2010	18:30	—	—	—	—	—	—	—	—	—
Quality-assurance samples												
Inorganic blank water sample		11/9/2006	14:00	—	—	—	5.46	—	8.31	—	—	—
Inorganic blank water sample		8/21/2009	12:20	—	—	—	< 5	—	7.96	—	—	—
Samples analyzed by Utah Department of Health Division of Epidemiology and Laboratory Services												
Mammoth Spring	10	7/18/1979	9:30	30	—	—	—	—	8	—	—	30
Mammoth Spring	10	8/31/2000	9:03	E 18	5 37	174	—	7.5	—	—	—	9.17
Mammoth Spring	10	9/6/2000	8:28	E 18	5 19	171	—	7.59	—	—	—	10.33
Mammoth Spring	10	7/16/2002	14:30	3.7	5 1	—	158	7.8	8.1	0	0.383	9.1
Mammoth Spring	10	8/26/2002	15:30	1.9	5 3	155	155	7.6	8.18	0	0.585	10.1
Mammoth Spring	10	9/23/2002	13:00	E 2.2	5 2	160	153	7.7	8.13	0	1.95	9.23
Mammoth Spring	10	10/23/2002	12:50	E 7.7	4 5	152	—	7.5	8.29	0	3.24	9.4
Mammoth Spring	10	1/23/2003	11:30	E 9	4 2	134	158	7.98	7.94	0	0.537	12.13
Mammoth Spring	10	2/26/2003	13:10	E 8.8	4 2	132	155	8.2	7.82	0	0.256	10.7
Mammoth Spring	10	3/28/2003	11:30	E 7.2	4	145	172	8.2	7.9	1.2	0.405	10.2
Mammoth Spring	10	4/28/2003	13:10	26.8	4 1	119	134	8.1	8.12	5.2	3.93	9.6
Mammoth Spring	10	5/28/2003	14:45	E 180	3 9	107	122	7.7	7.85	7.7	6.75	10
Mammoth Spring	10	6/23/2003	13:45	19	4.6	123	135	7.8	7.96	3.6	0.741	10
Mammoth Spring	10	9/5/2006	—	—	—	—	—	—	—	—	—	—
Mammoth Spring	10	8/21/2007	—	—	—	—	—	—	—	—	—	—

Table 2. Field measurements and results of chemical analyses for major ions, nutrients, and bacteria for selected groundwater and surface-water sites on the Markagunt Plateau, southwestern Utah.—Continued

[mm/dd/yyyy, month/day/year; hh mm, hour:minutes; ft³/s, cubic feet per second; °C, degrees Celsius; µS/cm, microsiemens per centimeter; NTU, nephelometric turbidity units; mg/L, milligrams per liter; ANC, acid neutralizing capacity; MPN, most probable number; ml, milliliters; E, estimated; —, not analyzed; <, less than; >, greater than]

Site name	Map ID Refer to Plate 1	Sample date (mm/dd/yyyy)	Time (hh:mm)	Dissolved oxygen, percent saturation	Alkalinity, field, mg/L as CaCO₃	Bicarbonate, mg/L as HCO₃	Carbonate, mg/L as CO₃	Alkalinity (ANC), lab, mg/L as CaCO₃	Hardness, mg/L as CaCO₃	Calcium, mg/L as Ca	Chloride, mg/L as Cl	Fluoride, mg/L as F
Mammoth Creek at Mammoth Spring	11	11/7/2006	16:10	—	112	133	1.3	125	120	34	2.1	E 0.064
Mammoth Creek rise pool	6	11/9/2006	13:30	—	116	141	0.4	128	123	35	2.2	< 0.1
		10/30/2008	15:00	—	—	—	—	—	—	—	—	—
Ephemeral spring	4	4/19/2007	21:30	—	106	128	0.4	121	118	34	2.3	E 0.08
Tommy Creek springs outflow	25	8/20/2007	20:15	—	90	109	< 1	91	85	20	1	0.19
Arch Spring	16	9/14/2007	16:10	—	157	192	< 1	160	157	42	1.6	0.15
Mammoth Creek springs	3	7/10/2008	10:00	—	106	128	0.3	111	108	31	1.4	0.12
Midway Creek at swallow holes	36	5/20/2009	19:40	—	116	142	< 1	112	119	42	4.2	< 0.08
Long Valley Creek at swallow holes	35	6/1/2009	15:45	—	164	200	< 1	158	163	50	0.89	E 0.064
West Asay Creek spring	26	8/21/2009	10:30	—	194	236	0.3	200	203	51	2.0	E 0.098
Duck Creek Lava Tube outflow	44	8/21/2009	13:45	—	132	161	0.1	132	134	37	2.4	E 0.065
		10/30/2008	11:15	—	—	—	—	—	—	—	—	—
Mammoth Creek at upper injection site	12	11/12/2009	17:00	—	108	132	0.1	116	110	32	2.1	E 0.067
Mammoth Spring	10	7/14/1954	—	—	—	—	—	—	—	—	—	—
Mammoth Spring	10	8/6/1954	—	—	—	—	—	—	70	—	2.5	—
Mammoth Spring	10	9/28/1968	—	—	—	—	—	—	70	20	2.5	—
Mammoth Spring	10	6/17/1989	16:30	—	82	100	< 1	—	82	19	1.1	0.3
Mammoth Spring	10	11/7/2006	14:45	—	—	—	—	80	76	21	0.7	0.2
Mammoth Spring	10	2/20/2007	18:30	—	75	91	0.3	85	81	22	1.4	E 0.096
Mammoth Spring	10	3/26/2007	16:00	—	—	—	—	—	—	—	—	—
Mammoth Spring	10	2/19/2008	15:00	—	—	—	—	—	—	—	—	—
Mammoth Spring	10	5/2/2008	13:00	—	—	—	—	—	—	—	—	—
Mammoth Spring	10	10/30/2008	14:15	—	66	80	< 1	67	64	19	2.4	E 0.12
Mammoth Spring	10	5/21/2009	11:30	—	—	—	—	—	—	—	—	—
Mammoth Spring	10	4/28/2010	18:30	—	68	83	0.2	67	71	22	1	E 0.056
Quality-assurance samples												
Inorganic blank water sample		11/9/2006	14:00	—	—	—	—	—	—	< 0.02	< 0.12	< 0.10
Inorganic blank water sample		8/21/2009	12:20	—	—	—	—	—	—	< 0.02	< 0.12	< 0.08
Samples analyzed by Utah Department of Health Division of Epidemiology and Laboratory Services												
Mammoth Spring	10	7/18/1979	9:30	—	—	—	—	—	—	—	—	—
Mammoth Spring	10	8/31/2000	9:03	84.2	—	—	—	—	—	—	—	—
Mammoth Spring	10	9/6/2000	8:28	94.4	—	—	—	—	—	—	—	—
Mammoth Spring	10	7/16/2002	14:30	94.4	—	96	0	79	73.8	18 9	< 3	—
Mammoth Spring	10	8/26/2002	15:30	98.8	—	99	0	81	80.8	21.7	< 3	—
Mammoth Spring	10	9/23/2002	13:00	95.3	—	101	0	83	79 1	20 5	< 10	—
Mammoth Spring	10	10/23/2002	12:50	96.6	—	184	0	151	141.7	40 3	< 10	—
Mammoth Spring	10	1/23/2003	11:30	121.4	—	97	0	79	76 9	19.8	< 10	—
Mammoth Spring	10	2/26/2003	13:10	99.6	—	95	0	78	75.7	19.8	< 10	—
Mammoth Spring	10	3/28/2003	11:30	102.9	—	106	0	87	87	23	< 10	—
Mammoth Spring	10	4/28/2003	13:10	95.5	—	79	0	64	70 1	18 9	< 10	—
Mammoth Spring	10	5/28/2003	14:45	99.9	—	73	0	60	65 2	20.4	10.4	—
Mammoth Spring	10	6/23/2003	13:45	93	—	85	0	69	67.6	18 9	< 10	—
Mammoth Spring	10	9/5/2006	—	—	—	—	—	—	—	—	—	—
Mammoth Spring	10	8/21/2007	—	—	—	—	—	—	—	—	—	—

Table 2. Field measurements and results of chemical analyses for major ions, nutrients, and bacteria for selected groundwater and surface-water sites on the Markagunt Plateau, southwestern Utah.—Continued

[mm/dd/yyyy, month/day/year; hh:mm, hour minutes; ft³/s, cubic feet per second; °C, degrees Celsius; µS/cm, microsiemens per centimeter; NTU, nephelometric turbidity units; mg/L, milligrams per liter; ANC, acid neutralizing capacity; MPN, most probable number; ml, milliliters; E, estimated; —, not analyzed; <, less than; >, greater than]

Site name	Map ID Refer to Plate 1	Sample date (mm/dd/yyyy)	Time (hh:mm)	Magnesium, mg/L as Mg	Potassium, mg/L as K	Silica, mg/L as SiO₂	Sodium, mg/L as Na	Sulfate, mg/L as SO₄	Dissolved-solids, residue on evaporation at 180°C, mg/L	Dissolved-solids, sum of constituents, mg/L	Ammonia, mg/L as N	Ammonia + organic nitrogen, mg/L as N
Mammoth Creek at Mammoth Spring	11	11/7/2006	16:10	8.5	1.2	19	2.9	1.5	—	E 144	< 0.02	E 0.088
Mammoth Creek rise pool	6	11/9/2006	13:30	8.7	1.2	19	2.9	1 5	—	147	< 0.02	< 0.1
		10/30/2008	15:00	—	—	—	—	—	—	—	—	—
Ephemeral spring	4	4/19/2007	21:30	7.8	1.4	17	2.6	1.4	—	140	< 0.02	—
Tommy Creek springs outflow	25	8/20/2007	20:15	8.4	1	16	3.8	1.6	—	108	< 0.02	—
Arch Spring	16	9/14/2007	16:10	13	0.63	12	1.7	1 3	—	170	< 0.02	—
Mammoth Creek springs	3	7/10/2008	10:00	7.3	1.4	21	2.9	1 3	—	130	—	—
Midway Creek at swallow holes	36	5/20/2009	19:40	3.1	0.5	6.5	3.3	0.93	139	131	—	—
Long Valley Creek at swallow holes	35	6/1/2009	15:45	9.0	0.37	7.2	0.97	0.75	175	E 168	—	—
West Asay Creek spring	26	8/21/2009	10:30	19	0.49	11	2	2	229	E 204	—	—
Duck Creek Lava Tube outflow	44	8/21/2009	13:45	10	0.61	9.3	1.7	0.8	135	E 142	< 0.02	—
		10/30/2008	11:15	—	—	—	—	—	—	—	—	—
Mammoth Creek at upper injection site	12	11/12/2009	17:00	7.4	1.3	20	2.9	1.6	130	E 132	—	—
Mammoth Spring	10	7/14/1954	—	—	—	—	—	—	—	—	—	—
Mammoth Spring	10	8/6/1954	—	—	—	—	—	—	—	—	—	—
Mammoth Spring	10	9/28/1968	—	4.7	—	20	—	3.6	103	—	—	—
Mammoth Spring	10	6/17/1989	16:30	8.3	1	18	3.4	3.5	104	104	—	—
Mammoth Spring	10	11/7/2006	14:45	5.8	1.1	19	3.2	1	—	101	< 0.01	—
Mammoth Spring	10	2/20/2007	18:30	6.3	1.2	20	3.6	1.7	—	109	< 0.02	< 0.10
Mammoth Spring	10	3/26/2007	16:00	—	—	—	—	—	—	—	—	—
Mammoth Spring	10	2/19/2008	15:00	—	—	—	—	—	—	—	—	—
Mammoth Spring	10	5/2/2008	13:00	—	—	—	—	—	—	—	—	—
Mammoth Spring	10	10/30/2008	14:15	4	0.81	13	2.5	1.4	97	86	E 0.01	—
Mammoth Spring	10	5/21/2009	11:30	—	—	—	—	—	—	—	—	—
Mammoth Spring	10	4/28/2010	18:30	4	0.83	13	2.4	1 2	91	E 86	—	—
Quality-assurance samples												
Inorganic blank water sample		11/9/2006	14:00	< 0.014	< 0.04	< 0 20	< 0.20	< 0.18	—	—	—	—
Inorganic blank water sample		8/21/2009	12:20	< 0.012	< 0.06	< 0 20	< 0 12	< 0.18	—	—	—	—
Samples analyzed by Utah Department of Health Division of Epidemiology and Laboratory Services												
Mammoth Spring	10	7/18/1979	9:30	—	—	—	—	—	98	—	< 0.1	—
Mammoth Spring	10	8/31/2000	9:03	—	—	—	—	—	—	—	< 0.05	—
Mammoth Spring	10	9/6/2000	8:28	—	—	—	—	—	—	—	< 0.05	—
Mammoth Spring	10	7/16/2002	14:30	6.47	1.24	—	3.91	< 20	110	—	—	—
Mammoth Spring	10	8/26/2002	15:30	6.47	1.26	—	3.83	< 20	100	—	—	—
Mammoth Spring	10	9/23/2002	13:00	6.8	1.31	—	3.65	< 20	106	—	—	—
Mammoth Spring	10	10/23/2002	12:50	10	2.61	—	9.49	< 20	210	—	—	—
Mammoth Spring	10	1/23/2003	11:30	6.69	1.31	—	4.16	< 20	104	—	—	—
Mammoth Spring	10	2/26/2003	13:10	6.38	1.21	—	3.7	< 20	110	—	—	—
Mammoth Spring	10	3/28/2003	11:30	7.2	1.2	—	4	< 20	112	—	—	—
Mammoth Spring	10	4/28/2003	13:10	5.57	1.35	—	3.07	< 20	84	—	—	—
Mammoth Spring	10	5/28/2003	14:45	3.48	< 1	—	2.38	< 20	78	—	—	—
Mammoth Spring	10	6/23/2003	13:45	4.97	1.22	—	3.22	< 20	88	—	—	—
Mammoth Spring	10	9/5/2006	—	—	—	—	—	—	—	—	—	—
Mammoth Spring	10	8/21/2007	—	—	—	—	—	—	—	—	—	—

[1] Second value is presumed to represent either replicate sample collected at site or sample taken at different location in spring

Table 2. Field measurements and results of chemical analyses for major ions, nutrients, and bacteria for selected groundwater and surface-water sites on the Markagunt Plateau, southwestern Utah.—Continued

[mm/dd/yyyy, month/day/year; hh mm, hour:minutes; ft³/s, cubic feet per second; °C, degrees Celsius; μS/cm, microsiemens per centimeter; NTU, nephelometric turbidity units; mg/L, milligrams per liter; ANC, acid neutralizing capacity; MPN, most probable number; ml, milliliters; E, estimated; —, not analyzed; <, less than; >, greater than]

Site name	Map ID Refer to Plate 1	Sample date (mm/dd/yyyy)	Time (hh:mm)	Nitrate plus nitrite, total, mg/L as N	Nitrate plus nitrite, mg/L as N	Nitrite, mg/L as N	Ortho-phosphate, mg/L as P	Total phosphorus, mg/L as N	Total phosphate, mg/L as P	Total coliform bacteria (MPN/100 ml)	Fecal coliform bacteria (MPN/100 ml)	Fecal Streptococcus group bacteria (MPN/100 ml)
Mammoth Creek at Mammoth Spring	11	11/7/2006	16:10	—	<0.06	<0.002	0.0078	<0.04	—	—	—	—
Mammoth Creek rise pool	6	11/9/2006	13:30	—	0.064	<0.002	0.013	<0.04	—	—	—	—
		10/30/2008	15:00	—	—	—	—	—	—	>300	7	—
Ephemeral spring	4	4/19/2007	21:30	—	E 0.05	—	0.015	—	—	—	—	—
Tommy Creek springs outflow	25	8/20/2007	20:15	—	0.3	—	0.045	—	—	—	—	—
Arch Spring	16	9/14/2007	16:10	—	0.42	—	0.015	—	—	—	—	—
Mammoth Creek springs	3	7/10/2008	10:00	—	—	—	—	—	—	—	—	—
Midway Creek at swallow holes	36	5/20/2009	19:40	—	—	—	—	—	—	—	—	—
Long Valley Creek at swallow holes	35	6/1/2009	15:45	—	—	—	—	—	—	—	—	—
West Asay Creek spring	26	8/21/2009	10:30	—	—	—	—	—	—	—	—	—
Duck Creek Lava Tube outflow	44	8/21/2009	13:45	—	0.1	—	0.009	—	—	—	—	—
		10/30/2008	11:15	—	—	—	—	—	—	180	0	—
Mammoth Creek at upper injection site	12	11/12/2009	17:00	—	<0.04	—	—	—	—	—	—	—
Mammoth Spring	10	7/14/1954	—	—	—	—	—	—	—	—	—	—
Mammoth Spring	10	8/6/1954	—	—	—	—	—	—	—	—	—	—
Mammoth Spring	10	9/28/1968	—	—	—	—	—	—	—	—	—	—
Mammoth Spring	10	6/17/1989	16:30	—	0.6	—	—	—	—	—	—	—
Mammoth Spring	10	11/7/2006	14:45	—	0.25	<0.01	0.04	—	—	—	—	—
Mammoth Spring	10	2/20/2007	18:30	—	0.33	<0.002	0.051	0.04	—	—	—	—
Mammoth Spring	10	3/26/2007	16:00	—	—	—	—	—	—	15	<1	—
Mammoth Spring	10	2/19/2008	15:00	—	—	—	—	—	—	22.2	1	—
Mammoth Spring	10	5/2/2008	13:00	—	—	—	—	—	—	71.7	<1	—
Mammoth Spring	10	10/30/2008	14:15	—	0.44	—	0.04	—	—	—	—	—
Mammoth Spring	10	5/21/2009	11:30	—	—	—	—	—	—	175	0	—
Mammoth Spring	10	4/28/2010	18:30	—	—	—	—	—	—	—	—	—
Quality-assurance samples												
Inorganic blank water sample		11/9/2006	14:00	—	—	—	—	—	—	27.1	<1	—
Inorganic blank water sample		8/21/2009	12:20	—	—	—	—	—	—	—	—	—
Samples analyzed by Utah Department of Health Division of Epidemiology and Laboratory Services												
Mammoth Spring	10	7/18/1979	9:30	0.15	—	—	—	—	0.05	40	<23	—
Mammoth Spring	10	8/31/2000	9:03	—	0.4	—	—	—	0.069	[1] 90/150	[1] 16/12	[1] 54/62
Mammoth Spring	10	9/6/2000	8:28	—	0.4	—	—	—	0.056	[1] 10/200	[1] 24/32	[1] <4/4
Mammoth Spring	10	7/16/2002	14:30	0.31	—	—	—	—	0.049	—	—	—
Mammoth Spring	10	8/26/2002	15:30	0.3	—	—	—	—	0.02446	—	—	—
Mammoth Spring	10	9/23/2002	13:00	0.37	—	—	—	—	0.02089	—	—	—
Mammoth Spring	10	10/23/2002	12:50	<0.1	—	—	—	—	0.025	—	—	—
Mammoth Spring	10	1/23/2003	11:30	0.31	—	—	—	—	0.059	—	—	—
Mammoth Spring	10	2/26/2003	13:10	0.29	—	—	—	—	0.04	—	—	—
Mammoth Spring	10	3/28/2003	11:30	0.25	—	—	—	—	0.047	—	—	—
Mammoth Spring	10	4/28/2003	13:10	0.45	—	—	—	—	0.023	—	—	—
Mammoth Spring	10	5/28/2003	14:45	0.85	—	—	—	—	0.027	—	—	—
Mammoth Spring	10	6/23/2003	13:45	0.46	—	—	—	—	0.021	—	—	—
Mammoth Spring	10	9/5/2006	—	—	—	—	—	—	—	73.8	0	—
Mammoth Spring	10	8/21/2007	—	—	—	—	—	—	—	73.8	12.4	—

Specific conductance of water from groundwater and surface-water sites at the time of sampling ranged from 137 to 364 μS/cm at 25°C, although specific conductance of water from most sites was less than 250 μS/cm (table 2). As with temperature, specific conductance can vary by individual spring and can vary with changes in temperature and discharge. On the basis of periodic measurements made during the study, specific conductance of water from Mammoth Creek rise pool was found to range from 307 μS/cm in mid-April 2008, prior to the start of runoff, to 191 μS/cm in late May 2009, during the peak snowmelt runoff, while specific conductance of water from Mammoth Creek, adjacent to the spring, ranged from 209 to 406 μS/cm (appendix 1). Similarly, specific conductance of water from Ephemeral spring ranged from 290 μS/cm in early August 2008 to 168 μS/cm in late May 2009, during the peak snowmelt runoff, while specific conductance of water from Mammoth Creek adjacent to the spring ranged from 142 to 413 μS/cm. The higher specific-conductance values for water from the creek represent water that discharges from other springs and spring-fed surface streams in the watershed upstream from these springs later in the year. Further, the substantially higher minimum values of specific conductance for water from the creek near Mammoth Creek rise pool (209 μS/cm) compared to values for water from the creek near Ephemeral spring (142 μS/cm) reflect the input of surface water that has higher conductance values discharging into Mammoth Creek between the two springs. The large variation in specific conductance coincident with a relatively small range in discharge of these springs compared to the wide range in conductance of water from Mammoth Creek adjacent to the springs, further indicates that a substantial amount of water discharging from the springs could originate from the creek.

Mammoth Creek springs (pl. 1, inset, site 3), which discharge into the streambed of Mammoth Creek, are likely the underflow (low stage) component of flow from Ephemeral spring, which is located only 100 ft away. Temperature and specific conductance of water from these springs are the same or very similar during periods of the year when snowmelt is insignificant and the flow in Mammoth Creek is dominated by springflow and overland runoff originating upstream. During these times, temperature and specific conductance of water from the creek are substantially higher than in water from the springs. During the snowmelt runoff period, however, subchannel (hyporheic) flow from Mammoth Creek appears to mix with the spring water discharging from the streambed, which results in the dilution, or reduction, of temperature and specific conductance.

The relatively high specific conductance of water from Long Valley Creek (289 μS/cm) and Midway Creek (230 μS/cm) that was measured during the snowmelt runoff period indicate substantial groundwater input from springs to these creeks. Expected values of specific conductance for snowmelt are less than 60 μS/cm, as measured in water losing to a sinkhole near The Craters area (pl. 1, site 39) during the same period. Although both of these creeks lose a substantial

amount of flow to the subsurface that eventually discharges at Mammoth Spring, as documented by dye-tracer tests (refer to discussion under "Dye-Tracer Studies"), the measured specific conductance was substantially greater than that measured at the spring during this period, indicating that most of the discharge from the spring is derived from snowmelt or other water with low-conductance values.

Temperature and specific conductance were periodically measured at several other springs on the plateau during the study for comparison purposes. These springs included Duck Creek Spring (site 53), Duck Creek Lava Tube outflow (site 44), Navajo Lake rise pool 2 (site 47), Navajo Lake Spring (site 58), Tommy Creek springs outflow (site 25), Cascade Spring (site 56), and Mammoth Creek springs (site 3) (appendix 1 and pl. 1). Duck Creek and Cascade Springs exhibited substantial variability in water temperature compared to other sites. Temperature of water from Duck Creek and Cascade Springs ranged from 6.3 to 11.7°C and from 8.7 to 14.2°C, respectively. Results of dye-tracer tests (refer to discussion under "Dye-Tracer Studies") have shown that a substantial portion of the discharge from these springs originates from subterranean diversion of outflow from Navajo Lake (pl. 1, site 49). During the study period, the dike impounding the lake was breached, and the outflow area below the dike was inundated. Consequently, the higher water temperatures of these springs are probably the result of warming of the shallow lake during the summer months.

Water from Duck Creek Lava Tube was also variable with respect to specific conductance and, to a lesser degree, temperature. The six measurements made during the study ranged from 148 to 308 μS/cm and from 6.0 to 8.5°C, respectively, while discharge ranged from about 0.25 ft³/s (100 gal/min) to 3.5 ft³/s (appendix 1). This variability results, in part, from the sources of water to the spring, which include inflow from Duck Creek Lake (pl. 1, site 52) and probably Duck Creek. In contrast, water from Navajo Lake Spring and Navajo Lake rise pool 2 showed little variation in temperature and specific conductance during periodic measurements, although differing from one another (appendix 1). The low variability in these water-quality measurements likely results from the source of recharge, which, in both cases, is probably diffuse infiltration of precipitation on the plateau directly north of the springs (pl. 1) combined with diffuse groundwater flow paths. Both of these springs typically discharge less than 1 ft³/s from the Claron Formation at about the same altitude along the north shore of Navajo Lake. Mean temperature and mean specific conductance of water from Navajo Lake Spring were 6.4°C and 346 μS/cm, in contrast to 3.5°C and 216 μS/cm for Navajo Lake rise pool 2. The reason(s) for the substantial difference in temperature and specific conductance between the springs is unknown.

Tommy Creek springs (pl. 1, sites 23 and 24), located in the lower part of the Tommy Creek drainage, which discharges to Mammoth Creek below Mammoth Spring, are most similar to Mammoth Spring with respect to temperature and specific conductance. Numerous periodic measurements of the combined flow of the springs showed a temperature of typically 5 to 6°C and specific conductance ranging between 160 and 190 µS/cm. Maximum discharge of the combined flow of the springs appears to be about 2.5 ft³/s before surface flow is initiated from snowmelt runoff higher in the drainage. Although very similar in chemistry to Mammoth Spring, and also likely discharging from the Claron, results of dye-tracer tests did not indicate a hydraulic connection between the springs. The likely source of water to Tommy Creek springs is from streamflow losses in tributaries to the main drainage upstream from the springs.

Hydrogen-ion activity (pH) in water from sampled sites and other selected locations across the plateau is shown in table 2 and in appendix 1. The pH of water from sampled springs ranged from 7.5 to 8.2, which is typical for groundwater discharging from carbonate aquifers, such as the Claron. The pH of water from Mammoth Spring ranged from 7.2 to 8.2 on the basis of measurements made during this study and those previously reported (table 2). The pH of water from surface-water sites sampled during the study ranged from 8.2 to 8.5. Loss of carbon dioxide (CO_2) downstream from springs discharging from the Claron along with aquatic and evapotranspiration processes in and adjacent to surface streams tend to generate slightly higher pHs than in groundwater discharging from springs.

Major Ions, Trace Elements, and Calculated Parameters

Water samples were collected from 12 groundwater (springs) and surface-water sites during the study period and analyzed for major ions, iron and manganese, alkalinity, and dissolved-solids concentration. In addition, one of the samples collected from Mammoth Spring was analyzed for a comprehensive suite of trace elements. Results of analyses for these sites, along with historical water-quality data for Mammoth Spring obtained from the U.S. EPA Storet database, are shown in tables 2 and 3. Major-ion analyses were used to determine the general chemistry of water discharging from various locations on the plateau, to help differentiate between water discharging from the Claron Formation and the overlying basalt, and to establish a baseline against which water quality can be compared over time. Samples from Mammoth Spring were collected at low, moderate, and high flows to compare variations in chemistry with changes in discharge.

Water from all sites can be classified as calcium bicarbonate on the basis of the predominance of these ions in terms of milliequivalents (fig. 10). Alkalinity values determined from field titrations ranged from 66 and 68 mg/L in water from Mammoth Spring during snowmelt runoff to 194 mg/L in water from West Asay Creek spring (table 2). Bicarbonate concentrations calculated from the alkalinity values for these sites were about 80 and 236 mg/L, respectively. Because the pH of water from sampled sites was generally less than 8.3, carbonate concentrations in water from all but one site

were less than 1 mg/L. All other ions were found in very low concentrations, and no constituents in water from any of the sites exceeded U.S. EPA or State of Utah primary or secondary standards (U.S. Environmental Protection Agency, 2009). In addition, on the basis of calcium and magnesium concentrations, calculated hardness values for all sites ranged from 64 to 203 mg/L, indicating generally moderately hard to hard water (Durfor and Becker, 1964), as would be expected in water discharging from carbonate bedrock.

Dissolved-solids concentrations in water from springs and surface-water sites generally were low across the plateau and ranged from only 91 mg/L in water from Mammoth Spring to 229 mg/L in water from West Asay Creek spring (table 2). Dissolved-solids concentration in water from most sites was less than 150 mg/L, which is reflected in specific-conductance values that were generally less than 300 µS/cm. The very low dissolved-solids concentrations in water from Mammoth Spring are unexpected when compared to water from other springs discharging from the Claron Formation. Reasons for the low concentrations could include rapid groundwater travel times in the aquifer, recharge through the overlying basalt, and, possibly, low concentrations of soil CO_2 during infiltration, all of which can affect dissolution of limestone units within the Claron. In comparison, the dissolved-solids concentration in water from West Asay Creek spring (pl. 1, site 26) was noticeably higher than in water from other springs discharging from the Claron. Concentrations of calcium, magnesium, and bicarbonate from this spring were also higher than those in water from other sites, whereas concentrations of other constituents were about the same (table 2). Water discharging from this spring could have a longer residence time in the Claron Formation or could be moving through zones within the Claron that are more soluble; both scenarios would result in higher concentrations of these constituents. Samples from Mammoth Creek at Mammoth Spring (site 11) and Mammoth Creek at upper injection site (site 12) were collected during baseflow conditions in November, when all flow in the creek is supplied by groundwater, and there is no dilution from snowmelt. As a result, water chemistry at these sites likely represents a composite of flow originating from multiple springs in the watershed upstream from these springs.

A comparison of the major-ion chemistry of water from all sites sampled other than Mammoth Spring shows that although some variance in calcium and magnesium occur, relative concentrations of the major ions are very similar among the sampled sites (fig. 10A). Samples collected from springs and surface water along Mammoth Creek upstream from Mammoth Spring had very similar calcium and magnesium concentrations compared to concentrations in samples from all other sites, which were more variable (fig. 10A), reflecting different sources. Further, the clustering of samples from Mammoth Creek, and Mammoth Creek rise pool and Ephemeral spring, is more evidence that water from these springs likely originates from the creek. On the basis of major-ion chemistry, groundwater movement through the basalt, such as that represented by outflow from Duck Creek Lava

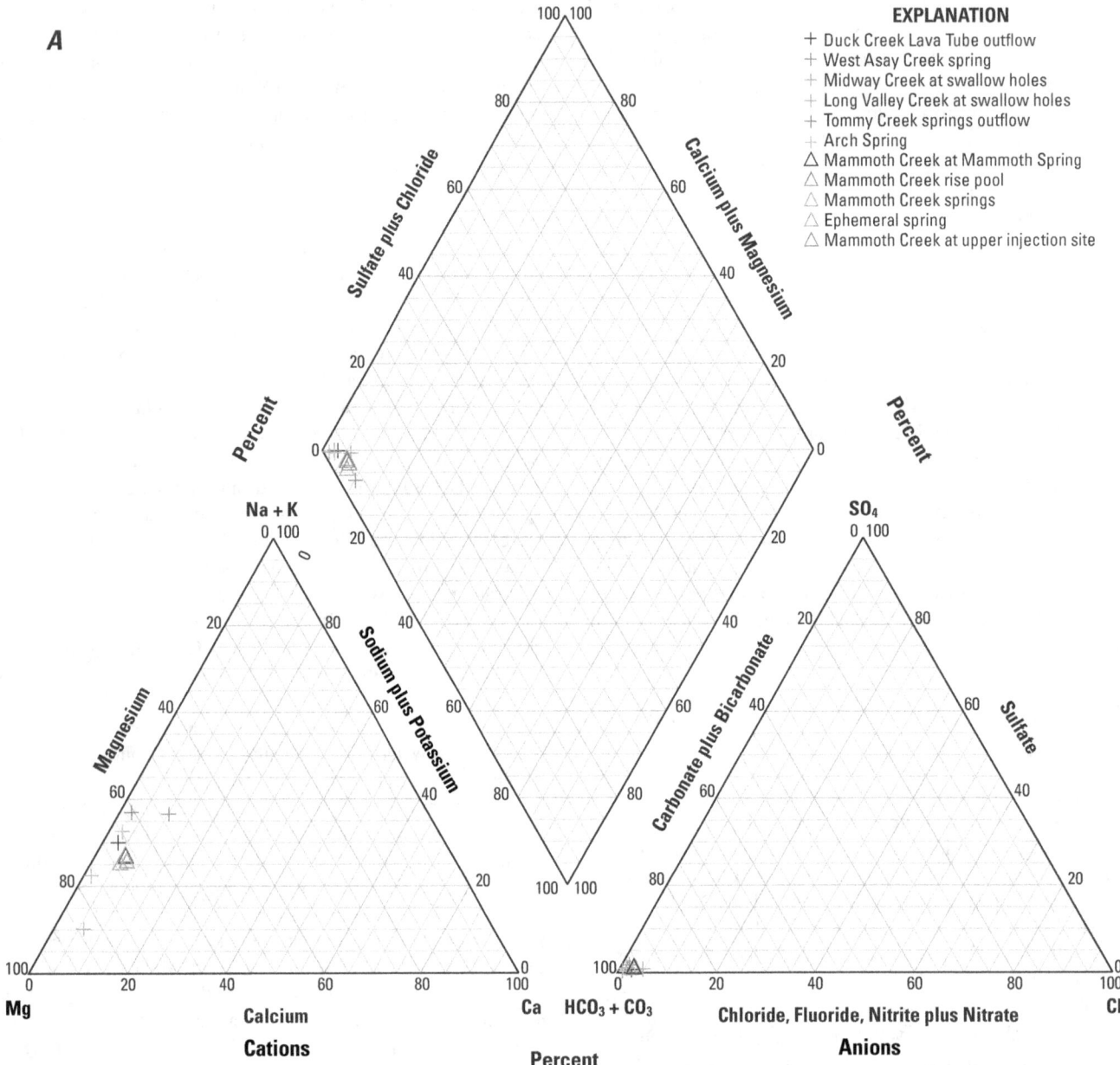

Figure 10.   Relative concentrations of major ions in **A**, selected ground and surface-water samples collected on the Markagunt Plateau, southwestern Utah, and **B**, samples collected from Mammoth Spring, 1968–2009.

Tube, could not be differentiated from water in the underlying limestone of the Claron Formation. These similarities are not surprising because groundwater flow is predominantly in the Claron, even where infiltration through the overlying basalt occurs. Because the basalt is relatively insoluble and relatively thin, residence time within the basalt is likely to be short as water moves downward along fractures, and the groundwater chemistry is dominated by water-rock interactions within the Claron Formation. Most groundwater samples were collected from springs that discharge from the Claron; thus, variations in dissolved-solids concentrations could be governed in large part, by groundwater residence time, which is determined

largely by source of recharge (sinkholes or diffuse infiltration) and flow path (conduit or matrix flow) within the aquifer.

Samples collected from Mammoth Spring during November 2006, May 2008, and May 2009 represent discharges of about 12, 88, and 200 ft³/s, respectively. The corresponding dissolved-solids concentrations for each of these water samples were 109, 97, and 91 mg/L, reflecting increasing dilution with an increase in the volume of snowmelt runoff in the aquifer. However, relative changes in major-ion concentrations with these changes in discharge were similar, except for magnesium, which showed a distinct trend of decreasing concentration with an increase in discharge, while calcium

B

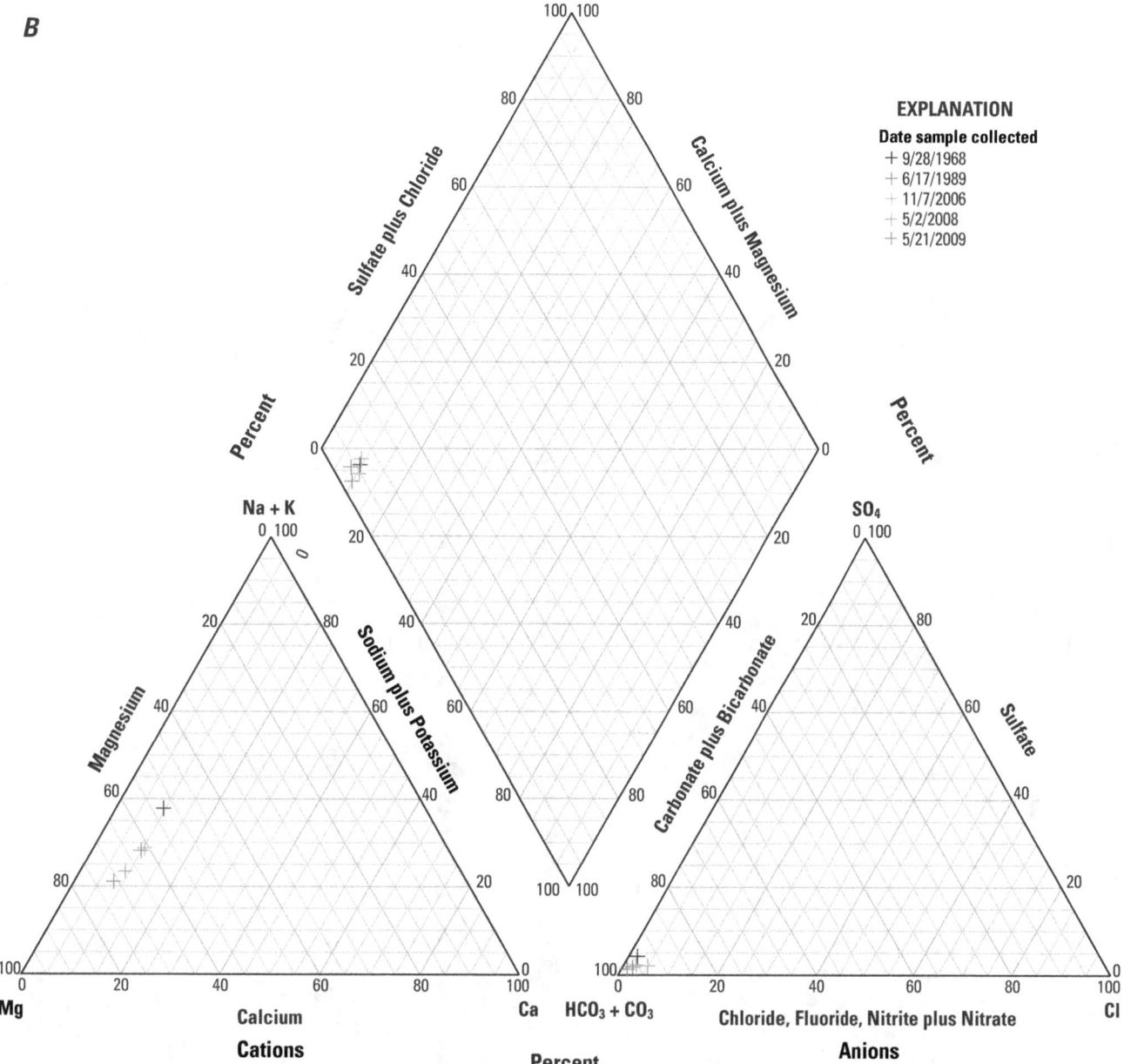

EXPLANATION

Date sample collected
+ 9/28/1968
+ 6/17/1989
+ 11/7/2006
+ 5/2/2008
+ 5/21/2009

Figure 10. Relative concentrations of major ions in *A*, selected ground and surface-water samples collected on the Markagunt Plateau, southwestern Utah, and *B*, samples collected from Mammoth Spring, 1968–2009.—Continued

concentration remained about the same (table 2 and fig. 10*B*). Results of analyses of samples collected prior to this study by the USGS, as well as analyses reported in the U.S. EPA Storet database (table 2), are very similar to the results of analyses of samples collected during this study and indicate that Mammoth Spring water chemistry has remained essentially the same over the last 50 years.

Results of analyses for trace-element concentrations in a sample collected from Mammoth Spring in November 2006 during baseflow conditions, when concentrations presumably would be highest and not affected by dilution from snowmelt, are presented in table 3. Concentrations of all constituents

analyzed were very low, or less than laboratory reporting levels, and did not exceed drinking-water standards (U.S. Environmental Protection Agency, 2009). Detectable concentrations of aluminum, titanium, and vanadium could be derived from infiltration through the basalt cap on the plateau. Arsenic, a common constituent of groundwater often associated with volcanic terrains, was detected at a concentration of 1.3 µg/L, which is well below the drinking-water standard of 10 µg/L. Strontium (78 µg/L) and barium (35 µg/L) concentrations likely originate from dissolution of limestone units within the Claron Formation. Iron concentrations in water from Mammoth Spring were quite variable during the study,

Table 3.   Results of trace-element analyses for selected groundwater and surface-water sites on the Markagunt Plateau, southwestern Utah.

[mm/dd/yyyy, month/day/year; hh:mm, hour:minutes; µg/L, micrograms per liter; mg/L, milligrams per liter; —, not analyzed; <, less than; E, estimated]

Site name	Map ID Refer to Plate 1	Sample date (mm/dd/yyyy)	Time (hh:mm)	Aluminum, µg/L as Al	Antimony, µg/L as Sb	Arsenic, µg/L as As	Barium, µg/L as Ba	Beryllium, µg/L as Be	Boron, µg/L as B	Bromide, mg/L as Br	Cadmium, µg/L as Ca	Chromium, µg/L as Cr	Cobalt, µg/L as Co	Copper, µg/L as Cu	Iron, µg/L as Fe	Lead, µg/L as Pb
Mammoth Creek at Mammoth Spring	11	11/7/2006	16:10	—	—	—	—	—	—	—	—	—	—	—	E 3.71	—
Mammoth Creek rise pool	6	11/9/2006	13:30	—	—	—	—	—	—	—	—	—	—	—	E 5.38	—
Ephemeral spring	4	4/19/2007	21:30	—	—	—	—	—	—	—	—	—	—	—	23	—
Tommy Creek springs outflow	25	8/20/2007	20:15	—	—	—	—	—	—	—	—	—	—	—	E 5.67	—
Arch Spring	16	9/14/2007	16:10	—	—	—	—	—	—	—	—	—	—	—	< 6	—
Mammoth Creek springs	3	7/10/2008	10:00	—	—	—	—	—	—	—	—	—	—	—	19.65	—
Midway Creek at swallow holes	36	5/20/2009	19:40	—	—	—	—	—	—	—	—	—	—	—	73	—
Long Valley Creek at swallow holes	35	6/1/2009	15:45	—	—	—	—	—	—	—	—	—	—	—	E 2.22	—
West Asay Creek spring	26	8/21/2009	10:30	—	—	—	—	—	—	—	—	—	—	—	< 4	—
Duck Creek Lava Tube outflow	44	8/21/2009	13:45	—	—	—	—	—	—	—	—	—	—	—	< 4	—
Mammoth Creek at upper injection site	12	11/12/2009	17:00	—	—	—	—	—	—	—	—	—	—	—	< 6	—
Mammoth Spring	10	6/17/1989	16:30	—	—	—	—	—	30	—	—	—	—	—	10	—
Mammoth Spring	10	11/7/2006	14:45	12	E 0.03	1.32	35	< 0.06	8.6	< 0.1	< 0.04	0.63	< 0.04	< 0.4	E 5.7	—
Mammoth Spring	10	5/2/2008	13:00	—	—	—	—	—	—	—	—	—	—	—	127	—
Mammoth Spring	10	5/21/2009	11:30	—	—	—	—	—	—	—	—	—	—	—	41	—
Quality-assurance samples																
Inorganic blank water sample		11/9/2006	14:00	—	—	—	—	—	—	—	—	—	—	—	< 6	—
Inorganic blank water sample		8/21/2009	12:20	—	—	—	—	—	—	—	—	—	—	—	< 4	—
Samples analyzed by Utah Department of Health Division of Epidemiology and Laboratory Services																
Mammoth Spring	10	7/16/2002	14:30	< 30	—	< 5	31	—	—	—	< 1	< 5	—	< 12	< 20	—
Mammoth Spring	10	10/23/2002	12:50	< 30	—	< 5	5.4	—	—	—	< 1	< 5	—	< 12	150	—
Mammoth Spring	10	1/23/2003	11:30	< 30	—	< 5	30.6	—	—	—	< 1	< 5	—	< 12	< 20	—
Mammoth Spring	10	4/28/2003	13:10	< 30	—	< 5	24.4	—	—	—	< 1	< 5	—	< 12	< 20	—

Table 3. Results of trace-element analyses for selected groundwater and surface-water sites on the Markagunt Plateau, southwestern Utah.—Continued

[mm/dd/yyyy, month/day/year; hh:mm, hour:minutes; μg/L, micrograms per liter; mg/L, milligrams per liter; —, not analyzed; <, less than; E, estimated]

Site name	Map ID Refer to Plate 1	Sample date (mm/dd/yyyy)	Time (hh:mm)	Manganese, μg/L as Mn	Mercury, μg/L as Hg	Molybdenum, μg/L as Mo	Nickel, μg/L as Ni	Selenium, μg/L as Se	Silver, μg/L as Ag	Strontium, μg/L as Sr	Thallium, μg/L as Tl	Titanium, μg/L as Ti	Tungsten, μg/L as W	Vanadium, μg/L as V	Zinc, μg/L as Zn
Mammoth Creek at Mammoth Spring	11	11/7/2006	16:10	0.309	—	—	—	—	—	—	—	—	—	—	—
Mammoth Creek rise pool	6	11/9/2006	13:30	E 0.13	—	—	—	—	—	—	—	—	—	—	—
Ephemeral spring	4	4/19/2007	21:30	0.55	—	—	—	—	—	—	—	—	—	—	—
Tommy Creek springs outflow	25	8/20/2007	20:15	0.41	—	—	—	—	—	—	—	—	—	—	—
Arch Spring	16	9/14/2007	16:10	<0.2	—	—	—	—	—	—	—	—	—	—	—
Mammoth Creek springs	3	7/10/2008	10:00	E 0.322	—	—	—	—	—	—	—	—	—	—	—
Midway Creek at swallow holes	36	5/20/2009	19:40	19	—	—	—	—	—	—	—	—	—	—	—
Long Valley Creek at swallow holes	35	6/1/2009	15:45	2.59	—	—	—	—	—	—	—	—	—	—	—
West Asay Creek spring	26	8/21/2009	10:30	E 0.16	—	—	—	—	—	—	—	—	—	—	—
Duck Creek Lava Tube outflow	44	8/21/2009	13:45	0.41	—	—	—	—	—	—	—	—	—	—	—
Mammoth Creek at upper injection site	12	11/12/2009	17:00	E 0.155	—	—	—	—	—	—	—	—	—	—	—
Mammoth Spring	10	6/17/1989	16:30	<1.0	—	—	—	<1	—	—	—	—	—	—	—
Mammoth Spring	10	11/7/2006	14:45	E 0.017	<0.01	0.43	0.13	0.11	E 0.05	78	<0.04	1.9	E 0.04	4.9	<0.6
Mammoth Spring	10	5/2/2008	13:00	1.7	—	—	—	—	—	—	—	—	—	—	—
Mammoth Spring	10	5/21/2009	11:30	0.76	—	—	—	—	—	—	—	—	—	—	—
Quality-assurance samples															
Inorganic blank water sample	10	11/9/2006	14:00	<0.2	—	—	—	—	—	—	—	—	—	—	—
Inorganic blank water sample		8/21/2009	12:20	<0.2	—	—	—	—	—	—	—	—	—	—	—
Samples analyzed by Utah Department of Health Division of Epidemiology and Laboratory Services															
Mammoth Spring	10	7/16/2002	14:30	<5	<0.2	—	—	<1	<2	—	—	—	—	<2	<30
Mammoth Spring	10	10/23/2002	12:50	184	<0.2	—	—	<1	<2	—	—	—	—	—	<30
Mammoth Spring	10	1/23/2003	11:30	<5	<0.2	—	—	<1	<2	—	—	—	—	—	<30
Mammoth Spring	10	4/28/2003	13:10	<5	<0.2	—	—	<1	<2	—	—	—	—	—	<30

ranging from an estimated 5.7 µg/L in the sample collected during November 2006 at baseflow, to 127 and 41 µg/L in the samples collected during May 2008 and May 2009, respectively, at high flow (table 3). Reasons for the high variability are unknown, but iron in the spring water is probably derived from infiltration through the basalt. The relatively high concentrations of iron (73 µg/L) and manganese (19 µg/L) in water from Midway Creek could be derived from older volcanic rocks within the surface drainage of the creek.

Nutrients and Bacteria

Results of analyses for nutrients are presented in table 2 for Mammoth Spring and for other groundwater and surface-water sites sampled during the study. Nutrients analyzed included ammonia, ammonia plus organic nitrogen, nitrate plus nitrite, nitrite, orthophosphate, and total phosphorus. Concentrations for all constituents were less than 0.5 mg/L, as would be expected in an alpine terrain where potential effects from agriculture or other anthropogenic activities are minimal, and likely represent natural background concentrations. Concentrations of nutrients reported in the U.S. EPA Storet database all were less than 1 mg/L as well (table 2). A sample collected during baseflow from Mammoth Creek at Mammoth Spring (site 11) also contained concentrations of nutrients that were near or less than laboratory reporting levels (table 2).

Total and fecal coliform bacteria samples were collected from Mammoth Spring on multiple occasions during the study. Total coliforms were consistently detected in water from the spring and ranged from 15 to more than 300 MPN per 100 mL of sample (table 2). Fecal coliforms, indicating a mammalian source, were generally less than 1 MPN per 100 mL in samples collected from the spring during this study. Results of analyses of samples reported in the U.S. EPA Storet database, however, were as high as 32 MPN of fecal coliforms per 100 mL of sample. In August 2007, sampling by the Utah Department of Environmental Quality resulted in 12.4 MPN of fecal coliforms per 100 mL of sample (Laurence Parker, written commun., 2007). No correlation appears to exist between total coliform count and discharge; the lowest and highest counts were from samples collected during baseflow conditions during the winter. The high variability in reported concentrations could be related to the location of the sampling point at the spring or sources of water to the spring at the time of sampling. Fecal source tracking methods that use extracted DNA from filtered water samples were used to differentiate human from all other mammalian bacteria in samples collected from Mammoth Spring and adjacent Mammoth Creek during baseflow conditions. Results of these analyses indicated the presence of fecal coliforms derived from general mammalian sources in both the spring and the creek, but fecal coliforms derived from human sources were detected only in the creek (Tricia Coakley, University of Kentucky, written commun., 2009). Although it is likely that coliform bacteria exist in the vicinity of the spring because they are ubiquitous in the natural environment, the repeated detection of these bacteria

in active flow from the spring outlet indicates that the bacteria likely originate from surface-water sources, such as Mammoth Creek, and are transported to the spring along high-permeability flow paths, such as dissolution-enlarged fractures.

Total coliform bacteria also were detected in a sample collected from the outflow of Duck Creek Lava Tube (site 44) in late October 2008, when discharge was less than 1 ft³/s. Results of this analysis indicated 180 colony producing units (CPU) or number of organisms per 100 mL of sample using the MF method. The outflow, which was the public drinking-water supply for the community of Duck Creek Village until 2008, was previously shown to be receiving water that loses from Duck Creek Lake, the probable source of the bacteria (Betsy Rieffenberger, U.S. Forest Service, internal memorandum dated August 19 and 21, 1975). In October 2008, total (greater than 300 CPU per 100 mL) and fecal (7 CPU per 100 ml) coliform bacteria also were detected in water from Mammoth Creek rise pool (pl. 1, inset, site 6), upstream from Mammoth Spring, further indicating that the rise pool could be hydraulically connected to the creek.

Stable and Radioisotopes, and Sulfur-35

Ten samples were collected from selected groundwater and surface-water sites on the Markagunt Plateau and analyzed for the stable isotopes of oxygen-18 (^{18}O) and deuterium (^{2}H). Three samples also were collected from Mammoth Spring at different discharge rates. These data are presented in table 4 and plotted in figure 11 relative to the global meteoric water line (GMWL). Variations in concentrations of ^{18}O and ^{2}H, expressed as delta (δ)^{18}O and delta (δ)^{2}H, in units of permil (per thousand) relative to Vienna Standard Mean Ocean Water (VSMOW), result from differences in ratios of these isotopes in precipitation, along with evaporative and altitude (temperature) effects (Clark and Fritz, 1997). As a result, δ^{18}O and δ^{2}H values in precipitation on the Markagunt Plateau are isotopically lighter (relatively more negative) than δ^{18}O and δ^{2}H values in precipitation at lower altitudes, such as in Cedar Valley to the west and about 4,000 ft lower, where evaporative effects can be much greater. On the Markagunt Plateau, variations in these isotopic values also can result from the time of year when sampling took place, the type of precipitation (rainfall or snowmelt), and differences in storm tracks. Precipitation on the plateau typically originates from the southwest and occasionally from the northwest; thus, isotopic ratios can vary from one storm event to another.

As shown in figure 11, all values, except one, plot close to or just above the GMWL and are relatively negative, largely because of the altitude effect. Isotopic values of water samples from Mammoth Spring (fig. 11, samples 10a, b, c) plot in a group about midway along the range of values from all sites, indicating a compositing or averaging of values representing water from different sources and altitudes. This averaging effect is also indicated by the relative positions of the isotopic values of samples from Long Valley Creek (sample 35), Midway Creek (sample 36), and Mammoth Creek at upper

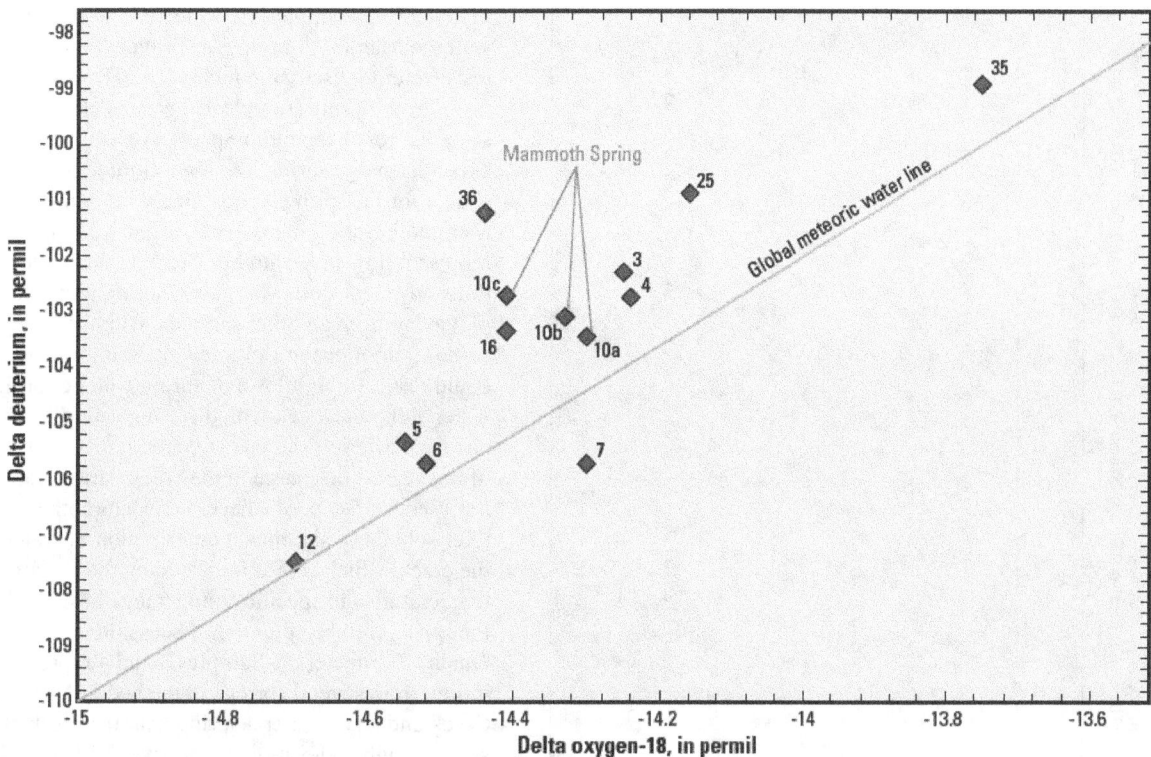

Site name	Map ID Refer to Plate 1	Sample date
Mammoth Creek springs	3	4/20/2007
Ephemeral spring	4	4/19/2007
Mammoth Creek above Ephemeral spring	5	7/10/2008
Mammoth Creek rise pool	6	11/9/2006
Mammoth Creek above Mammoth Creek rise pool	7	11/10/2006
Mammoth Spring	10a	11/7/2006
Mammoth Spring	10b	7/10/2008
Mammoth Spring	10c	5/21/2009
Mammoth Creek at upper injection site	12	11/12/2009
Arch Spring	16	9/14/2007
Tommy Creek springs	25	7/8/2008
Long Valley Creek at swallow holes	35	6/1/2009
Midway Creek at swallow holes	36	5/3/2008

Figure 11. Stable-isotope ratios of oxygen-18 and deuterium in water from selected groundwater and surface-water sites on the Markagunt Plateau, southwestern Utah.

Table 4.   Results of chemical analyses for stable and radioactive isotopes for selected groundwater and surface-water sites on the Markagunt Plateau, southwestern Utah.

[mm/dd/yyyy, month/day/year; hh:mm, hour:minutes; δ, delta; permil, per thousand; pCi/L, picocuries per liter; TU, tritium units; mBq/L, millibequerels per liter; —, not analyzed; ±, plus or minus]

Site name	Map ID Refer to Plate 1	Sample date (mm/dd/yyyy)	Time (hh:mm)	Deuterium (δ²H), permil	Oxygen (δ¹⁸O), permil	Tritium, pCi/L	Tritium, TU	Gross alpha, 30-day, Thorium-230, pCi/L	Gross alpha, 72-hour, Thorium-230, pCi/L	Gross beta, 30-day, Cesium-137, pCi/L	Gross beta, 72-hour, Cesium-137, pCi/L	Sulfur-35, mBq/L
Mammoth Creek rise pool	6	11/9/2006	13:30	-105.72	-14.52	—	—	—	—	—	—	—
Mammoth Creek above Mammoth Creek rise pool	7	11/10/2006	15:20	-105.73	-14.30	—	—	—	—	—	—	—
Ephemeral spring	4	4/19/2007	21:30	-102.74	-14.24	—	—	—	—	—	—	—
Mammoth Creek springs	3	4/20/2007	9:40	-102.28	-14.25	—	—	—	—	—	—	—
Arch Spring	16	9/14/2007	16:10	-103.36	-14.41	—	—	—	—	—	—	—
Midway Creek at swallow holes	36	5/3/2008	12:30	-101.23	-14.44	—	—	—	—	—	—	—
Tommy Creek springs outflow	25	7/8/2008	11:00	-100.88	-14.16	—	—	—	—	—	—	—
Mammoth Creek above Ephemeral spring	5	7/10/2008	13:00	-105.33	-14.55	—	—	—	—	—	—	—
Long Valley Creek at swallow holes	35	6/1/2009	15:45	-98.90	-13.75	—	—	—	—	—	—	—
Mammoth Creek at upper injection site	12	11/12/2009	17:00	-107.50	-14.70	—	—	—	—	—	—	—
Mammoth Spring	10	11/7/2006	14:45	-103.45	-14.30	28.4 ± 1.9	8.8	—	—	—	—	0.0 ± 0.3
Mammoth Spring	10	2/20/2007	18:30	—	—	—	—	-0.13 ± 0.60	0.41 ± 0.65	1.43 ± 0.62	0.49 ± 0.55	0.0 ± 0.3
Mammoth Spring	10	4/20/2007	15:00	—	—	—	—	—	—	—	—	4.0 ± 0.7
Mammoth Spring	10	7/10/2008	15:25	-103.08	-14.33	—	—	—	—	—	—	2.6 ± 0.4
Mammoth Spring	10	5/21/2009	14:40	-102.70	-14.41	—	—	—	—	—	—	—

injection site (sample 12), which are all sources of water to Mammoth Spring, as proven by dye-tracer tests (refer to discussion under "Dye-Tracer Studies"). Arch Spring (sample 16) groups closely with samples from Mammoth Spring. Although a relatively small discharge spring, the high-altitude recharge source for the spring is most likely the same as that for Mammoth Spring. Samples 5, 7, and 12 in figure 11 represent flow in Mammoth Creek at different locations upstream from Mammoth Spring and generally plot away from all other samples, with more negative values. Variations in isotopic values among the sites could result from inflows of spring water or surface water along the creek with differing isotopic ratios and (or) differing sources of water at different times of the year as discharge in the creek varies. Sample 6 represents a rise pool adjacent to Mammoth Creek that likely obtains a significant contribution of water from the creek, which is also indicated by the variations in temperature and specific conductance of water from the spring (refer to previous discussion under "Water-Quality Parameters"). Samples 3 and 4 represent inflow from several springs in the bed of Mammoth Creek and an ephemeral spring near the creek at a slightly higher elevation, respectively (pl. 1, inset). The similar isotopic ratio of water from these springs, along with other similarities in temperature and specific conductance discussed previously, indicate that the ephemeral spring is probably an overflow for the springs in the creek.

A sample for tritium was collected from Mammoth Spring during baseflow conditions when presumably most water from snowmelt runoff had moved through the aquifer and water from storage (longer residence time) discharged from the spring. Results of analysis of the sample collected in November 2006 indicated a concentration of about 28 pCi/L, or 8.8 tritium units, similar to present-day concentrations in precipitation for the region (Heilweil and others, 2006), and implying that water from aquifer storage is modern or relatively young.

Sulfur-35 was used to further refine age estimates of water from Mammoth Spring at four different times and discharge rates. Samples collected during baseflow conditions in November 2006 and February 2007 showed no statistically significant concentrations of sulfur-35 in the spring water (table 4). This indicated that a large component of the groundwater was more than about 1 year old and that runoff from the previous snowmelt runoff cycle had already passed through the aquifer. Water discharging from the spring during the winter months generally represents groundwater from storage, which has a longer residence time and should be devoid of sulfur-35. In contrast, water collected from the spring in April 2007 contained a significant concentration of sulfur-35 (about 4 mBq/L), indicating a short residence time likely representative of recharge that entered the aquifer during snowmelt runoff. Water samples collected from the spring in July 2008, during the recession of the snowmelt runoff cycle, also showed measurable amounts of sulfur-35 (about 2.6 mBq/L) that probably represented late spring snowmelt runoff that contained lower concentrations of sulfur-35 or a mixture of snowmelt runoff with an older component from storage.

Dye-Tracer Studies

The first regional hydrologic study on the Markagunt Plateau was carried out in the 1950s by the USGS to determine the discharge point of water from Navajo Lake losing to sinkholes and also the discharge point of water losing to Duck Creek Sinks. A localized study to determine the relation between Duck Creek Lake and water discharging from Duck Creek Lava Tube was carried out in 1975. The current study focused on the hydrology of the Mammoth Spring area and its relation to adjacent groundwater basins. Results of these investigations and several others are presented in table 5 and shown on plate 1.

Previous Investigations

Wilson and Thomas (1964) investigated groundwater movement along the southern edge of the Markagunt Plateau from 1954 to 1958, focusing on the hydrologic relations between Navajo Lake, and Cascade and Duck Creek Springs (pl. 1, sites 56 and 53). Navajo Lake is unusual in that most inflow originates from springs, which are located primarily along the north shore (pl. 1). In addition, basalt flows have blocked the natural surface-water outlet of the valley, which has resulted in subterranean piracy of outflow from Navajo Lake into Navajo Sinks (site 49). During a series of tests carried out in 1956, controlled releases of water from the lake into sinkholes below the dike impounding the lake produced increases in flow at Cascade Spring, 1.2 mi to the south, in about 1 hour, and at Duck Creek Spring, about 3 mi to the east, in about 12 hours. Results of dye-tracer tests, however, indicated groundwater travel times to these springs of about 8.5 and 53 hours, respectively (table 5). The considerably shorter travel times to these springs due to releases of water from the lake apparently result from propagation of a pressure wave through the aquifer, whereas travel times based on dye-tracing techniques represent actual movement of water through the aquifer. Apportionment of water to Cascade and Duck Creek Springs was calculated to be about 40 and 60 percent, respectively. A comparison of the amount of water lost into the sinkholes below the dike to the discharge of Cascade and Duck Creek Springs also indicated that these springs receive water from sources other than Navajo Lake as well.

Additional studies by Wilson and Thomas (1964) showed that Duck Creek Spring loses all flow into Duck Creek Sinks, about 2.5 mi east of the spring (pl. 1, sites 40, 41, and 45), which then resurges at Lower Asay Spring, about 6 mi farther

east (site 17). Flow increased at Lower Asay Spring in about 9 hours following a release of water to Duck Creek Sinks, while results of a dye-tracer test indicated an actual groundwater travel time of about 68 hours (table 5). Upper Asay Spring (site 18), located about 2,000 ft upstream from Lower Asay Spring, was not affected by the release of water into Duck Creek Sinks, nor was dye detected during the tracer test. Results of these investigations also showed that neither releases of water and subsequent increases in flow, or dye injected in Navajo Sinks or Duck Creek Sinks were detected at Mammoth Spring, about 8 mi to the north.

Bifurcation of the groundwater flow path, which presumably occurs in the vicinity of Navajo Sinks, results in discharge to different surface-water drainage basins (pl. 1). Cascade Spring discharges into the North Fork of the Virgin River, which lies within the Colorado River Basin, and Duck Creek Spring discharges into the Sevier River, which terminates in the Great Basin, through Duck Creek Sinks and Lower Asay Spring. Subterranean piracy of Navajo Lake outflow to Cascade Spring over time could result in increased flow to the Virgin River basin and a subsequent decrease in flow to the Sevier River basin (Wilson and Thomas, 1964).

Additional studies of the hydrology of the Duck Creek area were done in 1975 (Betsy Rieffenberger, U.S. Forest Service, internal memorandum dated August 19 and 21, 1975) to determine the source of fecal coliform bacteria to a water-supply spring discharging from Duck Creek Lava Tube, one of the longest lava tubes, about 2.27 mi, in the continental United States. Tracer injections along the northeast shore of Duck Creek Lake (pl. 1, site 52) and inside the lava tube, and monitoring inside of and at the outflow from the lava tube (site 44) showed that water from the lake discharged from springs inside the lava tube near the entrance and then discharged from the tube at the outflow spring. Groundwater travel time from inflow to the lava tube to the outflow point at the terminal end of the tube, approximately 7,500 ft, was less than 24 hours. Although these tracer tests showed a hydraulic connection between Duck Creek Lake and outflow from the lava tube, streamflow inside the lava tube upstream (up tube) from the input point from the lake indicates an additional component of flow that is likely derived from infiltration through the overburden above the lava tube, as well as from possible losing streams on the surface in the vicinity of the lava tube.

Investigations from 2007–2011

During 2007 to 2009 and in 2011, seven dye injections were made in the central and southwestern parts of the Markagunt Plateau to better understand groundwater flow directions, rate of movement, and relations between Mammoth and other groundwater basins. Four tracer tests were carried out in the southwestern part of the plateau in the Midway Valley area, two tests took place in the Mammoth Creek drainage upstream from Mammoth Spring, and an additional test was completed in a tributary in the upper reaches of Tommy Creek (table 5). Dye was recovered in all but the test from the Tommy Creek drainage.

Table 5.   Results of dye-tracer tests for selected springs on the Markagunt Plateau, southwestern Utah.

[mm/dd/yyyy, month/day/year; hh:mm, hour:minutes; lb(s), pound(s); ft³/s, cubic feet per second; h, hours; ND, no data; E, estimated; gal/min, gallons per minute; d, days; <, less than]

Site name	Tracer injection					
	Altitude (feet)	Date (mm/dd/yyyy)	Time (hh:mm)	Type of tracer	Amount of tracer	Discharge of injection point
Asay - Cascade Springs basin						
Navajo Sinks[1]	9,020	8/12/1954	8:20	Fluorescein dye	0.5 lb	3 2 ft³/s
Duck Creek Sinks[1]	8,370	8/24/1954	15:40	Fluorescein dye	1 5 lbs	ND
Duck Creek Lake[2]	8,540	8/4/1975	10:35	Fluorescein dye	ND	ND
Duck Creek Lava Tube[2]	[3] 8,535	8/4/1975	10:25	Fluorescein dye	ND	ND
Mammoth Spring basin						
Tributary to upper Tommy Creek	8,620	5/10/2007	14:45	Fluorescein dye	5 lbs	E 20–25 gal/min
Midway Creek at swallow holes	9,620	6/11/2008	19:00	Fluorescein dye	14 lbs	E 200–250 gal/min
Mammoth Creek at campground	8,170	10/30/2008	16:40	Rhodamine WT	1 liter	E 200 gal/min
Stream sink near The Craters	10,020	5/20/2009	16:30	Fluorescein dye	20 lbs	E 100 gal/min

Table 5.   Results of dye-tracer tests for selected springs on the Markagunt Plateau, southwestern Utah.—Continued

[mm/dd/yyyy, month/day/year; hh:mm, hour:minutes; lb(s), pound(s); ft³/s, cubic feet per second; h, hours; ND, no data; E, estimated; gal/min, gallons per minute; d, days; <, less than]

Site name	Altitude (feet)	Discharge of recovery point (at time of injection)	Date (mm/dd/yyyy)	Time (hh:mm)	Travel time to tracer recovery (first detection)	Linear distance (feet)	Vertical distance (feet)	Other sites monitored
				Tracer recovery				
Asay - Cascade Springs basin								
Cascade Spring	8,760	2.2 ft³/s	8/12/1954	16:50	8.5 h	6,400	160	
Duck Creek Spring	8,550	ND	8/14/1954	13:20	53 h	18,500	475	
Lower Asay Spring	7,140	ND	8/28/1954	12:00	68 h	36,000	1,260	West Asay Creek Mammoth Spring Upper Asay Spring
Duck Creek Lava Tube	[3] 8,535	ND	8/5/1975	10:30	24 h	E 500	5	
Duck Creek Lava Tube outflow	8,400	ND	8/5/1975	10:30	24 h	7,400	135	
Mammoth Spring basin								
No recovery[4]	ND	ND	ND	ND	ND	ND	ND	Tommy Creek springs outflow Mammoth Spring Mammoth Spring at confluence Mammoth Creek above highway Mammoth Creek rise pool
Mammoth Spring	8,125	64 ft³/s	7/8/2008	13:50	[5] 27 d	44,500	1,500	Mammoth Creek rise pool
Mammoth Spring at confluence	8,120		7/8/2008	13:15				Ephemeral spring
Mammoth Creek above highway	7,790		7/8/2008	10:30				Mammoth Creek at Mammoth Sp. Navajo Lake rise pool 1 Navajo Lake rise pool 2 Navajo Lake Spring Ashdown Creek above confluence Three Creeks at Larson Ranch Duck Creek Spring Duck Creek Spring outflow Tommy Creek springs outflow
Mammoth Spring	8,125	7.7 ft³/s	1/19/2009	12:20	[5] 81 d	1,250	45	Tommy Creek springs outflow
Mammoth Creek at Mammoth Sp.	8,125	< 5 gal/min	1/19/2009	15:00	[5] 81 d	1,250	45	Mammoth Creek above highway
Mammoth Spring at confluence	8,125		5/27/2009	17:45	[5] 7 d	47,500	1,900	Mammoth Creek rise pool
Mammoth Spring	8,125	207 ft³/s	6/26/2009	12:20				Ephemeral spring
Mammoth Creek above highway	7,790		6/26/2009	17:15				Mammoth Creek at Mammoth Sp. Navajo Lake rise pool 2 Navajo Lake Spring Ashdown Creek above confluence Shooting Star Creek Three Creeks at Larson Ranch Deep Creek at Taylor Ranch Duck Creek Spring Duck Creek Spring outflow Tommy Creek springs outflow Cascade Spring

Table 5.   Results of dye-tracer tests for selected springs on the Markagunt Plateau, southwestern Utah.—Continued

[mm/dd/yyyy, month/day/year; hh:mm, hour:minutes; lb(s), pound(s); ft³/s, cubic feet per second; h, hours; ND, no data; E, estimated; gal/min, gallons per minute; d, days; <, less than]

| | Tracer injection | | | | | |
Site name	Altitude (feet)	Date (mm/dd/yyyy)	Time (hh:mm)	Type of tracer	Amount of tracer	Discharge of injection point
Mammoth Spring basin—Continued						
Long Valley Creek at swallow holes	9,740	6/1/2009	12:45	Tinopal CBS-X	25 lbs	E 2 ft³/s
Mammoth Creek at upper injection site	8,420	11/12/2009	16:00	Fluorescein dye	3 5 lbs	E 100 gal/min
Ashdown Creek basin						
Stream sink along Highway 148	10,180	6/26/2011	21:00	Fluorescein dye	23 lbs	E 2 ft³/s

[1] Data from Wilson and Thomas (1964)

[2] Letter dated August 19 and 21, 1975, Betsy Rieffenberger, Dixie National Forest

[3] Altitude in lava tube is projected

[4] Dye may have discharged downstream from Mammoth Spring in Mammoth Creek

[5] Dye recovered on activated charcoal; maximum travel time

[6] Dye observed visually in Mammoth Spring by Mammoth Creek resident

[7] Dye observed visually in Coal Creek by Forest Service personnel

[8] Altitude, linear, and vertical distance based on presumed discharge from Arch Spring in Cedar Breaks

Tommy Creek springs discharge year-round in the lower part of the Tommy Creek drainage at two principal sites (pl. 1, sites 23 and 24) and at several additional locations during the snowmelt runoff period. The main drainage of Tommy Creek is dry above the springs for most of the year. During snowmelt runoff, surface flow is present in the entire drainage and mixes with the flow of water from the springs, which are located adjacent to the creek and discharge into Mammoth Creek about 1.5 mi downstream from Mammoth Spring. Dye injected into a losing stream in a tributary in the upper reaches of Tommy Creek (site 30) was not recovered at any monitored sites, including Mammoth Spring and Tommy Creek springs outflow (site 25), which were considered the most likely discharge locations. The dye could have discharged into Mammoth Creek downstream from its confluence with Mammoth Spring or at another unmonitored location.

Dye-tracer tests in the Midway Creek and Long Valley Creek areas and in an area near (south of) The Craters were carried out in 2008–09 (pl. 1; table 5). Altitudes of the dye-injection points ranged from about 9,600 to 10,000 ft. In June 2008, fluorescein dye was injected into swallow holes in the channel of Midway Creek along highway 14 (site 36) and recovered at all monitored locations in Mammoth Spring as well as downstream in Mammoth Creek above the highway (site 19). Because Mammoth Spring discharges from multiple outlets during the snowmelt runoff period, monitoring of the spring was done at five sites: two at the main springhead, one downstream from the springhead and just upstream from the Troll monitor, one in a rise pool on the south side of and discharging into the main channel, and one downstream at the confluence of the spring run with Mammoth Creek. All of these sites were monitored to determine their relation to one another hydrologically. Dye was not detected at other sites monitored during the test, which included the two rise pools upstream from Mammoth Spring (pl. 1, inset, sites 4 and 6), Mammoth Creek at Mammoth Spring (site 11), Tommy Creek springs outflow (site 25), several springs along the north shore of Navajo Lake (sites 46 and 47), Navajo Lake Spring

Table 5. Results of dye-tracer tests for selected springs on the Markagunt Plateau, southwestern Utah.—Continued

[mm/dd/yyyy, month/day/year; hh:mm, hour:minutes; lb(s), pound(s); ft³/s, cubic feet per second; h, hours; ND, no data; E, estimated; gal/min, gallons per minute; d, days; <, less than]

Site name	Altitude (feet)	Discharge of recovery point (at time of injection)	Date (mm/dd/yyyy)	Time (hh:mm)	Travel time to tracer recovery (first detection)	Linear distance (feet)	Vertical distance (feet)	Other sites monitored
				Tracer recovery				
Mammoth Spring basin—Continued								
Mammoth Spring	8,120	77 ft³/s	6/26/2009	12:00	[5] 25 d	44,000	1,620	Tommy Creek springs outflow
Mammoth Spring at confluence	8,125		6/26/2009	11:25				Cascade Spring
								Ashdown Creek above confluence
								Navajo Lake rise pool 2
								Duck Creek Spring
								Duck Creek Spring outflow
								Mammoth Creek rise pool
Mammoth Spring[6]	8,125	4.8 ft³/s	11/13/2009	ND	< 1 d	6,800	320	Tommy Creek springs outflow
Mammoth Spring at confluence	8,120		11/13/2009	ND	< 1 d			
Mammoth Creek above Tommy Cr.	7,840		11/13/2009	ND	< 1 d			
Mammoth Creek above highway	7,790		11/13/2009	ND	< 1 d			
Ashdown Creek basin								
Ashdown Creek below confluence[7]	[8] 9,080	E 20–25 ft³/s	6/27/2011	ND	< 11 h	[8] 11,850	[8] 1,100	Mammoth Spring
								Navajo Lake Spring
								Deep Creek at Taylor Ranch
								Three Creeks at Larson Ranch
								Ephemeral spring
								Mammoth Creek rise pool
								Mammoth Spring at confluence
								Tommy Creek springs outflow
								Mammoth Creek above highway
								Duck Creek Spring

(site 58), Duck Creek Spring (sites 53 and 54), Three Creeks at Larson Ranch, a spring-fed creek in the southwestern part of the study area (site 60), and Ashdown Creek above its confluence with Shooting Star Creek (site 14), which includes the outflow from Arch Spring (site 16). Maximum groundwater travel time to Mammoth Spring from Midway Creek was shown to be 27 days on the basis of the detection of dye on the first set of detectors pulled from the spring. Because this approach represents an integration of dye concentration over time, actual travel time was likely substantially less than this (refer to results of the dye-tracer test near The Craters in next paragraph).

In May and June 2009, two tracer injections were made, one from a stream sink near The Craters (site 39) and one from swallow holes in the streambed of Long Valley Creek (site 35) (table 5). Fluorescein dye injected into the stream sink was recovered at all monitored locations in Mammoth Spring (see previous paragraph) as well as downstream in Mammoth Creek above the highway (site 19). Dye was not detected at other sites monitored during the test, which included all of

the sites monitored during the tracer test from Midway Creek plus Shooting Star Creek (site 15), Cascade Spring (site 57), and Deep Creek at Taylor Ranch (site 59). Optical brightener injected into swallow holes in Long Valley Creek also was recovered at all monitored locations in Mammoth Spring and was not detected at other selected sites monitored during the test (table 5). Maximum groundwater travel time to Mammoth Spring from the stream sink near The Craters was found to be 7 days on the basis of the detection of dye on the first set of detectors pulled from the spring. Maximum groundwater travel time to Mammoth Spring from Long Valley Creek was 25 days, which also was based on the first set of detectors pulled from the spring. Although the Long Valley Creek tracer test was carried out during the snowmelt runoff period only 10 days after the injection into the stream sink near The Craters, and from a similar distance, discharge of the spring was substantially less, implying slower groundwater velocities (table 5). Nonetheless, actual groundwater travel time from Long Valley Creek was likely substantially less than 25 days.

Tracer tests from Midway Creek and Long Valley Creek, and from the stream sink near The Craters, indicate a convergence of flow paths to Mammoth Spring, which is typical of flow paths documented in other karst regions (Mull and others, 1988; Spangler, 2001). Although represented as straight-line vectors, groundwater flow paths to Mammoth Spring are likely developed along tortuous routes that are influenced by regional structure and stratigraphy. As a result, actual flow-path distances can be 30 to as much as 50 percent longer than that shown by the straight-line representations (Mull and others, 1988). By using results of the tracer test from the stream sink near The Craters, and assuming a maximum travel time of 7 days, average groundwater velocity to the spring would have been 6,000 to 7,000 ft/d or more, a value typically encountered in karst areas where conduit flow predominates (Worthington, 2007). This high groundwater velocity is enhanced by the relatively steep hydraulic gradient (4 percent) between the injection point and the spring.

Water loses directly into the Claron Formation in Midway Creek and loses through channel-fill deposits in Long Valley Creek, which then funnel water into the underlying Claron. North of this area in the central part of the plateau, however, basaltic lava flows overlie the Claron Formation and infiltration of rain and snowmelt takes place directly through the basalt and into the underlying formation. In some areas, groundwater probably moves laterally within the basalt for some distance before encountering fractures or other vertical pathways into the Claron. In some instances, groundwater also apparently moves laterally along contact zones, such as between the base of lava flows and the original valley floors, to discharge from the toe of lava flows (pl. 1, sites 31 and 37). Because Mammoth Spring also discharges from the Claron Formation, the principal groundwater flow paths are likely developed in this unit along fractures, faults, and bedding planes (structural dip of the Claron) that have been enlarged by dissolution. In some areas of the plateau, northeast-trending faults are present along which large sinkholes have developed (Moore and others, 2004), indicating the influence of faults on the movement of groundwater within the Claron Formation. Development of these sinkholes and numerous others in the central part of the plateau undoubtedly has resulted in more focused recharge into the basalt and further enhanced dissolution of the underlying Claron.

To help define the direction of groundwater movement along the western margin of the plateau, a tracer test was carried out in June 2011 from a stream sink located west of highway 148 (pl. 1, site 38; table 5). Fluorescein dye injected at this location was visually observed in Coal Creek in Cedar Canyon, approximately 3 mi downstream from Cedar Breaks National Monument (the study area boundary), and less than 11 hours from the time of injection (Chris Butler, Dixie National Forest, written commun., 2011). Dye also was visually observed in Ashdown Creek above its confluence with Coal Creek but was not observed in Coal Creek upstream from this junction, indicating that the dye likely originated from springs in Cedar Breaks. Subsequent analysis of detectors placed downstream from the confluence of Ashdown Creek and Shooting Star Creek (site 13) verified that the source of the dye originated from Cedars Breaks and possibly from Arch Spring (site 16), about 3.6 mi upstream from the confluence. Arch Spring has the largest known spring discharge in the Monument and is a principal contributor to flow in Ashdown Creek (pl. 1). Results of this test showed that groundwater moves north from the vicinity of site 38 along the western margin of the plateau into Cedar Breaks, likely along mapped, north-trending faults, before presumably discharging from the Claron Formation approximately 1,100 ft lower in elevation. Dye was not recovered from detectors placed at other monitored sites southwest of the plateau or to the southeast along Navajo Lake and at Duck Creek Spring (table 5). More significantly, dye was not detected at Mammoth Spring, indicating a groundwater divide between the injection site and the stream sink near The Craters (site 39) about 1.6 mi to the east (pl. 1).

Additionally, two tracer tests were carried out in losing reaches along Mammoth Creek upstream from Mammoth Spring (pl. 1, inset; table 5). The confluence of the spring with Mammoth Creek occurs about 200 ft below the springhead. After the onset of snowmelt runoff and during the summer months, flow is present in the channel of Mammoth Creek to the confluence. During the fall, flow in the channel normally begins to recede upstream from the confluence to a Forest Service campground (pl. 1, inset, site 2), where it stabilizes from input of flow primarily from Mammoth Creek rise pool (site 6). Flow in the channel upstream from the rise pool is generally perennial but minimal, and provided by streams that enter the channel on the south side and by the flow of Mammoth Creek springs (site 3), which rise in the bed of the creek near Ephemeral spring (site 4). Upstream from Ephemeral spring, flow in the channel is again minimal during the late fall and is provided by streams that enter the channel on the south side from the Dead Lake area (pl. 1, inset). Upstream from these inputs, the channel is typically dry for about 3,000 ft, at which point all perennial flow in the channel provided by streams from the north and south sides infiltrates the channel deposits (pl. 1, inset).

Dye injected in the channel of Mammoth Creek about 1,250 ft upstream from Mammoth Spring at the campground (site 2) in October 2008 was detected at Mammoth Spring and also in a small amount of flow discharging from the channel near the confluence (site 11). Water movement from the dye-injection site to the confluence was likely through sub-channel routes (hyporheic flow) within the bouldery stream deposits. Dye from a second tracer test in November 2009, about 1.25 mi upstream from the spring, where all flow infiltrates the channel (site 12), also was detected at the spring and observed visually in outflow from the spring less than 1 day after injection (Ann Harris, Mammoth Creek resident, oral commun., 2010), indicating a groundwater travel time of more than a mile per day. Flow infiltrating the channel deposits presumably moves into the underlying Claron Formation where dissolution-enlarged fractures conduct water to the spring. Results of these tracer tests showed that once flow in Mammoth Creek begins to recede late in the year, all flow infiltrating the channel likely discharges at the spring. Because the discharge of Mammoth Spring typically is less than 10 ft³/s during this time of year, water from the creek makes up a substantial portion of baseflow. During full channel flow in the spring and summer months, an unknown but probably relatively minor portion of flow in the creek discharges at the spring.

Groundwater Basin Delineation

On the basis of tracer studies completed during 2007–09 and in 2011, the recharge area or groundwater basin for Mammoth Spring is interpreted to include the area within the watershed of upper Mammoth Creek, about 40.5 mi², as well as an area southwest of the spring and outside of the watershed in which the spring is located, estimated to be at least 25 mi² (pl. 1). North of the Mammoth Creek watershed boundary, groundwater movement is probably toward Blue Spring (site 1). South of Mammoth Creek, water discharging from the spring originates primarily from precipitation (mainly snowmelt) that infiltrates directly through the basalt and through focused recharge points, particularly sinkholes and swallow holes along streambeds. This includes Midway, Long, and Sage Valleys, and the Horse Pasture, Hancock Peak, and Red Desert areas (pl. 1). North of Mammoth Creek, and to some degree, south of the creek, surface flow appears to predominate, and focused points of recharge, such as the numerous sinkholes developed on the plateau south of the Mammoth Creek watershed boundary, are distinctly absent. Absence of these features appears to be in large part related to differences in the geology north and south of the creek. North of the creek, older Tertiary-age volcanic rocks and Quaternary-age surficial materials overlie the Claron, and are less permeable than the basaltic lava flows that dominate the geology on the south side of the creek. This can act to inhibit downward movement of water into the Claron where dissolution can take place and sinkholes can develop. Nonetheless, water that does not run off infiltrates to the water table and probably moves downgradient into Mammoth Creek, where it then moves downstream to losing reaches. Although the area outside and south of the Mammoth Creek watershed could be substantially smaller than the area within the watershed, results of dye-tracer tests and the abundance of focused points of recharge in this area indicate that most of the recharge to Mammoth Spring likely originates from this part of the plateau.

Relations between the Tommy Creek drainage and Mammoth Spring are unclear, and additional dye-tracer tests are needed to resolve groundwater-surface-water relations in this area. Tommy Creek springs also discharge from the Claron Formation, and specific conductance of water from the springs is very similar to that of Mammoth Spring. Nonetheless, although the springs were monitored during all tracer studies, the absence of dye in outflow from the springs indicates that the recharge area for the springs could be localized, probably within the drainage upstream from where the springs are located, and not hydraulically connected to the Mammoth

Spring basin. The western boundary of the Mammoth Spring groundwater basin probably is defined, in part, by the watershed boundary of Mammoth Creek, which is also the escarpment of Cedars Breaks National Monument (pl. 1). In the southeastern part of Cedar Breaks, however, this boundary could be farther east because the recharge area for Arch Spring probably lies, at least in part, east-southeast of the spring, on the basis of the orientation of the principal conduit from which the spring discharges. Results of a dye-tracer test from site 38 (pl. 1) in 2011 also indicate that groundwater moves from the south into Cedar Breaks, forming a divide between this area and groundwater moving to the northeast to Mammoth Spring. Discharge characteristics of Arch Spring, which is the likely discharge point for groundwater along this part of the western margin of the Markagunt Plateau, are unknown. Thus, the size of the groundwater basin supplying the spring cannot be accurately determined in relation to the Mammoth Spring basin.

The southwestern boundary of the recharge area for Mammoth Spring probably lies southwest of Midway Valley near the Pink Cliffs (pl. 1). The southern boundary of the groundwater basin is not accurately defined but tracer studies in the Navajo Lake watershed, previously discussed, along with a series of relatively low-discharge springs located along the north shore of Navajo Lake, indicate that the boundary between the two basins could lie in the Deer Valley area (pl. 1). Recharge to the springs along the north shore of the lake likely originates from the area directly north of the lake, where the Claron Formation is exposed at the surface. The lower reaches of Midway Creek southeast of its junction with Deer Valley (pl. 1) probably lose water to Duck Creek and (or) Cascade Springs, or possibly to one of the springs along the north shore of Navajo Lake. The southeastern boundary of the Mammoth Spring groundwater basin, including the Tippets Valley area (pl. 1), is poorly established, and additional dye-tracer tests are needed to more accurately define directions of groundwater movement in this area and hydrologic boundaries between the Mammoth Spring, and Duck Creek, Cascade, and Asay Springs groundwater basins.

Hydrologic Relations

Relations among precipitation, discharge, and water quality indicate that Mammoth Spring is capable of responding rapidly to recharge events from distant areas within the groundwater basin of the spring and that physical characteristics such as temperature, specific conductance, and turbidity also change with fluctuations in discharge. These responses and associated changes, along with the results of dye-tracer tests, can be used for approximating groundwater travel times from different areas in the basin, which then can be used to evaluate the potential effects of anthropogenic activities.

Relation between Precipitation and Discharge

Precipitation is measured hourly at the Natural Resources Conservation Service (NRCS) Snotel site in Midway Valley, which lies within the inferred boundary of the groundwater basin of Mammoth Spring at an altitude of 9,800 ft, about 10 mi southwest of the spring (pl. 1). Total precipitation for the 2007 water year (October 1, 2006, to September 30, 2007) was 27.8 in. Total precipitation for the 2008 and 2009 water years was 32.0 and 33.2 in., respectively. For comparison, mean annual precipitation for the period of 1971–2000 was about 37 in. (*http://www.wcc.nrcs.usda.gov/nwcc/ site?sitenum=626&state=ut*).

Discharge (stage) of Mammoth Spring was measured on a continuous basis (1- and 2-hour intervals) from November 2006 to December 2009. Discharge of the spring, expressed as daily mean values, is shown in figure 12 for the 3-year period. The peak daily mean discharge in 2008 was 199 ft³/s on May 19 (instantaneous peak discharge of 218 ft³/s), following a 5-week rise from a baseflow of about 6 ft³/s. In 2009, peak daily mean discharge was 224 ft³/s on May 12–13 (instantaneous peak discharge of about 240 ft³/s), following a 4-week rise from a baseflow of about 6 ft³/s. Response of Mammoth Spring to snowmelt runoff was substantially different in 2007 than in 2008 and 2009. In 2007, maximum discharge of Mammoth Spring was bimodal, rather than the more typical single peak observed in 2008 and 2009 (fig. 12). The initial peak daily mean discharge was about 54 ft³/s on April 11, a month earlier than in 2008 and 2009, followed by a second almost identical peak discharge of about 56 ft³/s (instantaneous peak

discharge of 59 ft³/s) almost a month later on May 4–5. The bimodal peak in flow resulted from a late spring snowstorm on April 22–23 that along with cooler temperatures, substantially decreased the discharge of the spring from its initial peak. The second peak of similar magnitude occurred after warmer temperatures resumed and induced snowmelt again. Time to peak discharge from prior low-flow conditions was 10 to 11 days in both cases. In addition, during the winter and spring months prior to the principal snowmelt runoff period, numerous smaller increases (spikes) in discharge were recorded (fig. 12). The erratic response of the spring during this period is attributed to warmer periods that resulted in premature melting and subsequent increases in discharge of the spring. This early loss of snow, combined with less snowpack than in 2007–08 and 2008–09, resulted in a smaller peak flow during late spring, when there normally would be a single, larger peak flow.

A comparison of Mammoth Spring discharge and precipitation from November 2006 to November 2007 is shown in figure 13. Response of Mammoth Spring to rainfall events on the plateau is shown in late July and in early and late September, and can be used as a surrogate for assessing groundwater travel times to the spring. The most dramatic response occurred on September 24, in response to 1.1 in. of rain recorded at the Midway Valley precipitation station on September 22 (fig. 13). Daily mean discharge of the spring increased from about 9.4 ft³/s on the day of the event to 30 ft³/s on the 23rd, and to 46 ft³/s on the 24–25th, before rapidly decreasing to 16 ft³/s by the 27th. Because temperature and specific conductance decreased with the increase in flow (refer to next section on "Relation between Water Quality

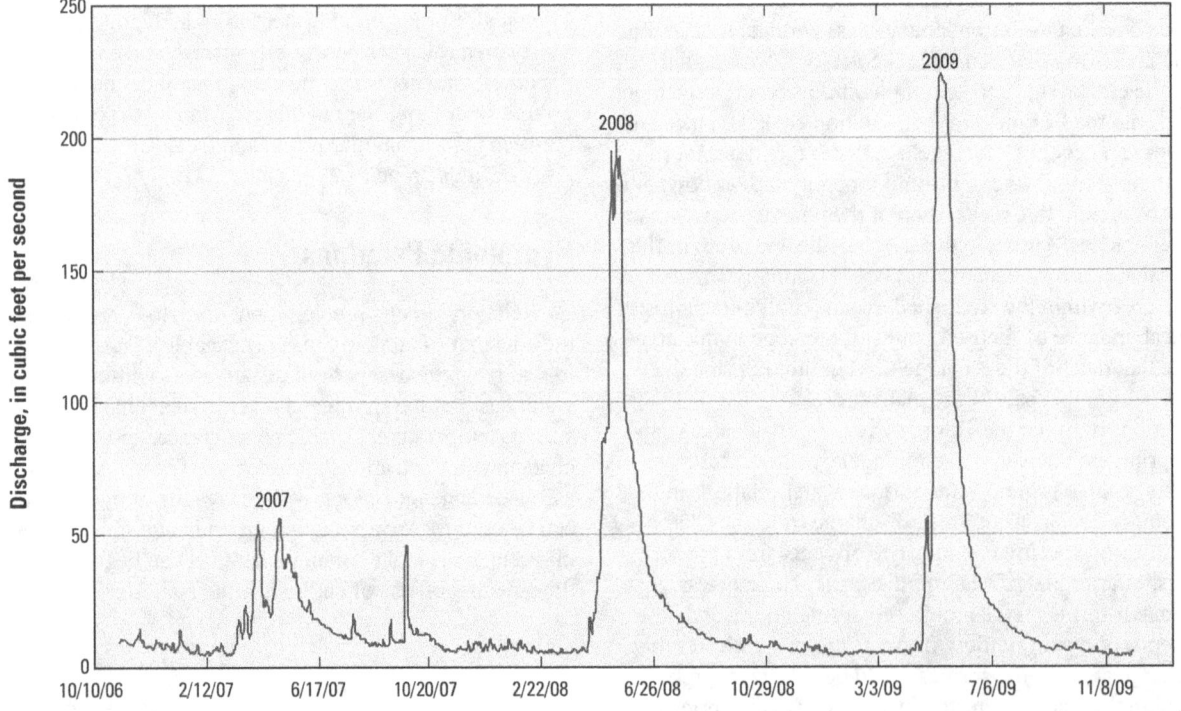

Figure 12.   Daily mean discharge for Mammoth Spring, Markagunt Plateau, southwestern Utah, November 2006 to December 2009.

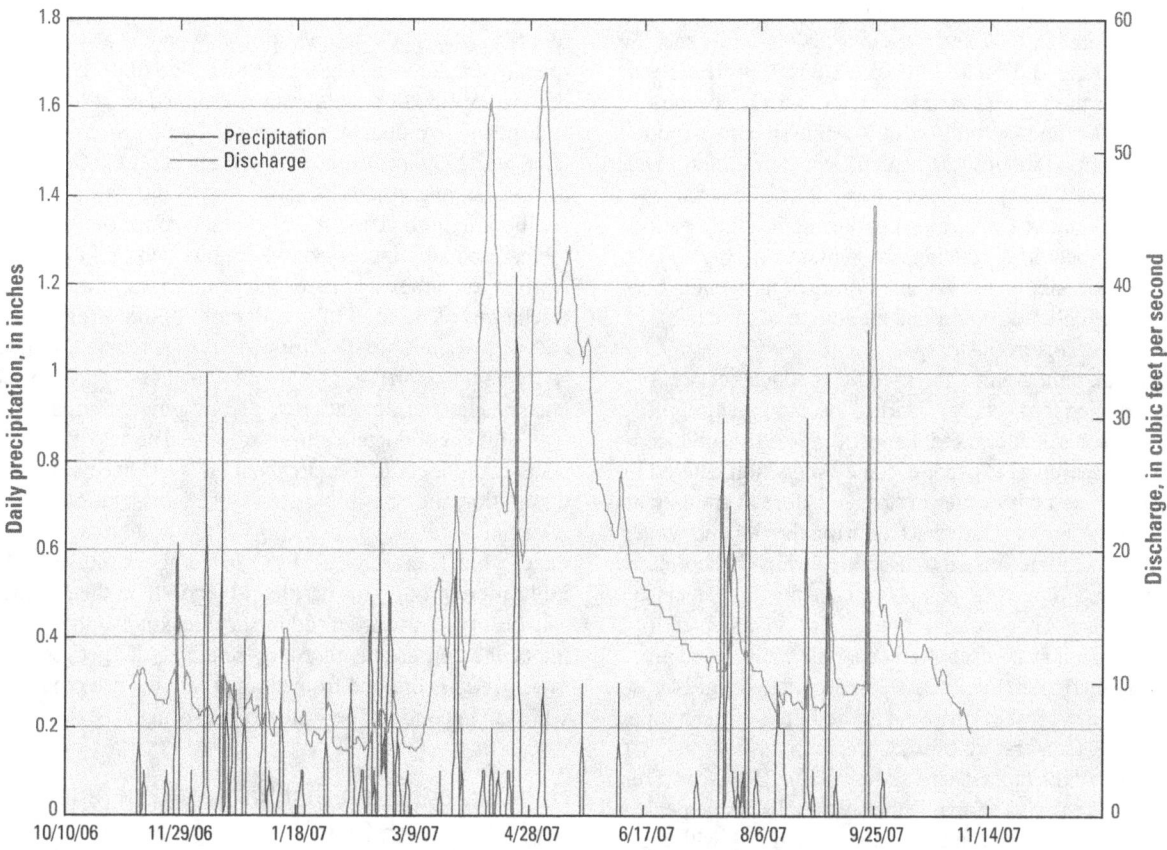

Figure 13.   Comparison of Mammoth Spring daily mean discharge with precipitation in Midway Valley, Markagunt Plateau, southwestern Utah, November 2006 to November 2007.

and Discharge"), the increase in discharge of the spring is not attributed to the propagation of a pressure wave through the aquifer but instead to the actual movement of water from the surface to the spring along dissolution-enlarged fractures or other high-permeability flow paths. This was substantiated by observations of increased turbidity in the spring water during this period (Chris Butler, Dixie National Forest, oral commun., 2007). This rapid response in discharge likely originated from focused points of recharge, such as sinkholes and swallow holes in streambeds on the plateau or, possibly, from losing reaches along Mammoth Creek. Similar responses to rainfall events also occurred on July 24–26, when an increase in the daily mean discharge of about 9 ft³/s, nearly doubling flow, was noted following 0.9 and 0.7 in. of rain on July 21 and 24, respectively, and on September 6, when an increase of 7 ft³/s in 1 day was noted after 1 in. of rain fell on the plateau the previous day (fig. 13).

Increase in flow of Mammoth Spring in response to rainfall events can be variable and is related to the intensity and duration of the event, extent of the area of precipitation, and antecedent soil-moisture conditions at the time of the event, among other factors. Evapotranspiration can play a significant role in the amount of recharge to the aquifer as well, particularly during the summer months. The lack of response in discharge

from the spring after a significant rainfall event (1.6 in.) on August 1 could be attributed to one or a combination of these factors. During the summer and fall of 2008, several 0.5-in. daily precipitation events were noted; however, minimal to no response was recorded at the spring. The largest amount of rain, 1.0 in., fell on July 28, but response at the spring was only about 3 ft³/s. Small responses (about 2 ft³/s) also were recorded at the spring in the first part of October in response to snowmelt (1-in. water equivalent) on the plateau. Response to snowmelt events is typically slower than to rainfall events because the water is gradually released to the aquifer. No significant precipitation events were recorded on the Markagunt Plateau during the summer and fall of 2009.

Relation between Water Quality and Discharge

Temperature and specific conductance of water from Mammoth Spring were measured on a continuous basis (1- and 2-hour intervals) from November 2006 to December 2009. Relations between these measurements and discharge are shown in figures 14, 15, and 16. On the basis of daily mean values, temperature of water from the spring ranged from 3.8 to 5.3°C between November 2006 and November 2007, while specific conductance ranged from 129 to 168 µS/cm.

Temperature of water from the spring ranged from 3.6 to 5°C between November 2007 and November 2008, while specific conductance ranged from 132 to 203 μS/cm. Temperature of water from the spring ranged from 3.4 to 5°C between November 2008 and December 2009, while specific conductance ranged from about 112 to 198 μS/cm. The overall range in temperature of the spring water, regardless of discharge, only varied by about 1.5°C annually during the study period. The range in specific conductance was more variable and appears to be related to the volume or rate of movement of snowmelt through the aquifer and the length of time that groundwater drains from storage.

In general, temperature and specific conductance are inversely related to discharge. During the snowmelt runoff period, as discharge increased, temperature and specific conductance decreased as low-conductance snowmelt entered the aquifer and mixed with water in storage before eventually discharging at the spring. Conversely, during the fall and winter months, discharge decreased to baseflow, which is dominated by water from storage that is higher in specific conductance and temperature. These relations have been documented in other alpine karst systems as well (Spangler, 2001) and are particularly evident in the 2006–07 hydrographs (fig. 14). During the period just prior to and at the initial discharge peak in April 2007, temperature of water decreased by about 0.3°C, and specific conductance decreased by about 35 μS/cm. This inverse relation is also shown at the second discharge peak and during the interval between the peaks, particularly with regard to specific conductance. Similar relations can be observed during rainfall events, where cooler, low-conductance precipitation mixes with water in conduits and fractures within the surrounding matrix. During the September 26, 2007, rainfall event, temperature of the spring water dropped 0.7°C, and specific conductance decreased by about 30 μS/cm. The magnitude or amount of change in these values resulting from these "dilution events" can be highly variable.

Temperature of the spring water gradually increased after snowmelt runoff and discharge waned in all 3 years monitored (figs. 14, 15, and 16) and reached peak values each year in August before decreasing to background (baseflow) values during the winter months. During the summer months, water temperatures were a degree or more higher than during the winter months. Several factors could account for the higher water temperatures after the runoff period. Surface-water temperatures are warmest during this period, in response to increasing air temperature, and are typically three times warmer than the spring water (appendix 1). Influx of surface waters to the aquifer, particularly Mammoth Creek, along with rapid groundwater travel times as documented by dye-tracer tests, could result in an increase of the spring water temperature above the average baseflow temperature. In addition, warming of the spring outflow between the discharge area and the downstream temperature sensor also could be a potential factor contributing to the increased temperature.

Temperature and specific conductance decreased as spring discharge increased during the snowmelt runoff period in

2007–08 and 2008–09 (figs. 15 and 16). During the peak runoff in May 2008, temperature dropped by about 0.5°C, and specific conductance decreased by about 70 μS/cm. During the peak runoff in 2009, temperature dropped by at least 0.6°C, and specific conductance decreased by 86 μS/cm. In both 2008 and 2009, the onset of snowmelt runoff coincided with the largest magnitude of change in specific conductance. In addition, in both years, specific conductance of water from the spring gradually increased in February and peaked in mid-April at its highest value for the year, prior to the increase in discharge (rising limb of hydrograph) of the spring (figs. 15 and 16), when it rapidly dropped off as snowmelt in the spring water began to arrive. This gradual increase in specific conductance is interpreted to result from gravity drainage of water with higher conductance from storage. These storage components could include fractures that have not been enlarged by dissolution and possibly the epikarst, a dissolution-enhanced zone likely developed at the top of the unsaturated zone in the Claron. The source of water in storage is mostly diffuse infiltration through the basalt and regolith on the plateau. The observed upward trend in specific conductance appears to occur as progressively older (relatively longer residence time) water is drained from storage into the principal conduits, which then conduct the water to the spring.

Relation between Discharge of Mammoth Spring and Mammoth Creek

Discharge of the combined flow of Mammoth Spring and Mammoth Creek is measured at USGS streamgaging station 10173450, "Mammoth Creek above west Hatch ditch, near Hatch, Utah," located approximately 8.5 mi downstream from Mammoth Spring. Flow at the gaging station includes inflow from other tributaries between the spring and the gage, such as Tommy Creek, but these inflows are usually minor compared to the flows of the spring and Mammoth Creek. As a result, the hydrograph at the gage generally represents flow from these two sources. Historically, the minimum daily mean flow at the gage has been less than 5 ft³/s on numerous occasions (*http://waterdata.usgs.gov/ut/nwis/uv?site_no=10173450*) and generally is so during the winter months, when virtually all flow originates from the spring. The relation between the discharge of Mammoth Spring and the combined flow of the spring and the creek, expressed as daily mean discharge, is shown in figure 17 for the November 2006 to December 2009 monitoring period. The hydrograph shows a strong correlation between spring discharge and total streamflow for this period and indicates that response of the spring and runoff in the creek to snowmelt are almost simultaneous. Although the hydrographs for the spring and total flow in the creek show a steep concurrent rising limb, the separation between the spring and total streamflow becomes more apparent on the recession limb, when discharge from the spring decreases at a slightly faster rate. Ratios between the peak runoff for total streamflow and springflow show that discharge from Mammoth Spring

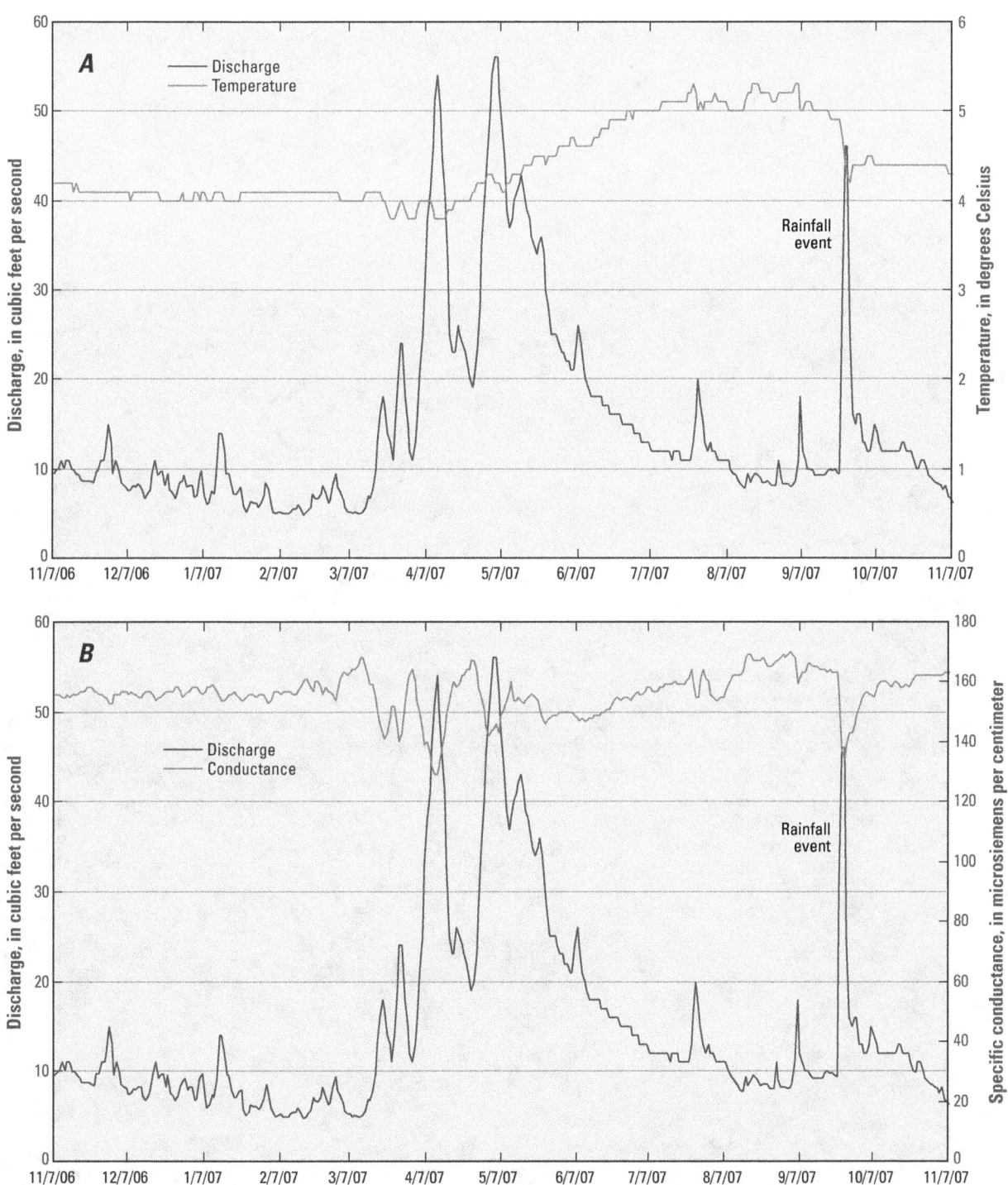

Figure 14.   Relation between discharge of Mammoth Spring, November 2006 to November 2007, and **A**, water temperature, and **B**, specific conductance.

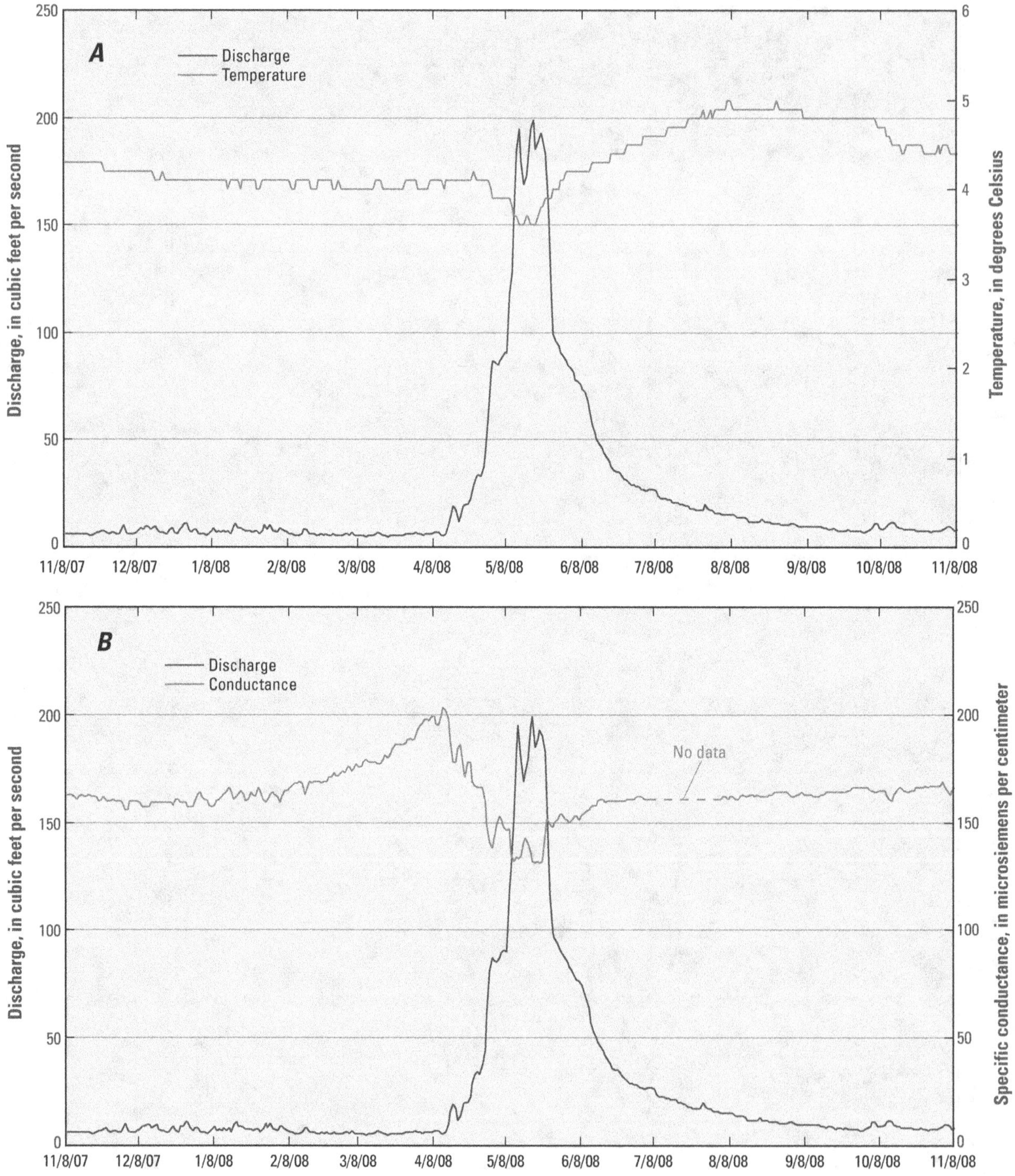

Figure 15.   Relation between discharge of Mammoth Spring, November 2007 to November 2008, and **A**, water temperature, and **B**, specific conductance.

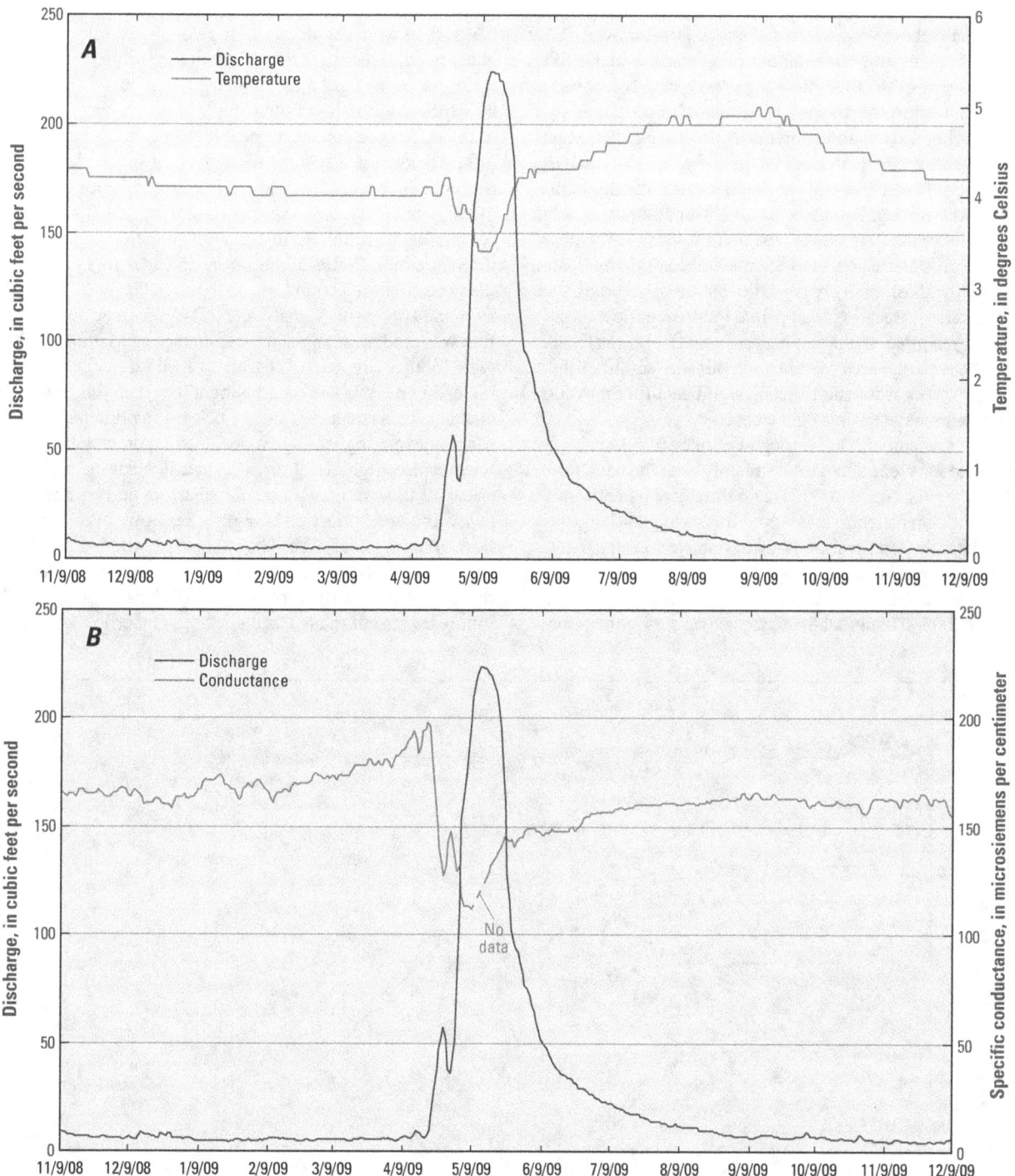

Figure 16.　Relation between discharge of Mammoth Spring, November 2008 to December 2009, and **A**, water temperature, and **B**, specific conductance.

accounted for about 54 percent of the total streamflow in 2007 and 2008 and about 46 percent of the total streamflow in 2009 (fig. 17). At other times of year, this percentage varied, and a substantial part of the time, flow from the spring composed most or all of the total streamflow measured at the gage, such as during the fall and winter of 2007–08 (fig. 17). Most rainfall events on the plateau result in an increase in discharge of Mammoth Creek, but may or may not affect the discharge of Mammoth Spring. Response to the 1.1-in. rainfall event of late September 2007, however, resulted in a peak daily mean flow of 47 ft³/s at the gage and 46 ft³/s at the spring, indicating that virtually all of the increase in flow at the gage originated from the spring. Because no apparent increase in flow of the creek was recorded, the precipitation event likely was localized to the recharge area of the spring outside (south) of the Mammoth Creek watershed, and no significant runoff occurred within the drainage basin of the creek.

During December 2007 through February 2008, springflow appeared to have exceeded total streamflow at the gage for much of the time (fig. 17). This undoubtedly resulted from ice forming downstream, away from the spring discharge point, where the temperature is always above freezing. An additional source of streamflow loss could be through the bouldery streambed. In contrast, during much of the fall and winter of 2006–07 and 2008–09, as well as during the fall of 2009, when no flow occurred in Mammoth Creek above the confluence with Mammoth Spring, total streamflow recorded at the gage exceeded discharge from the spring by a factor of more than two (fig. 17). Although some of this difference can be attributed to inflow from tributaries above the gage, most of the difference appears to be due to channel (streambed) inflow below the discharge measurement site for Mammoth Spring, which was also below the springflow monitoring (Troll) site during the study period. Measurements made on November 12, 2009, during baseflow conditions, indicated a discharge of 5.25 ft³/s at the spring and 14.2 ft³/s just below the Forest Service boundary, or about 2,600 ft downstream from the spring (pl. 1, inset, site 22). A measurement of 14.0 ft³/s was made just above the confluence of Mammoth Creek with Tommy Creek (site 20), about 1 mi farther downstream, indicating no measurable additional inflow in this reach of the channel. This streambed inflow below the principal outlet of Mammoth Spring was again documented on October 2, 2010, when a discharge of 9.9 ft³/s was measured at the spring and about 19 ft³/s were measured downstream at the same location below the Forest Service boundary, resulting in an increase in flow of more than 9 ft³/s. The streambed inflow downstream from Mammoth Spring could be additional groundwater that represents an underflow component of the spring, or could be inflow that is not related to the spring. Underflow components

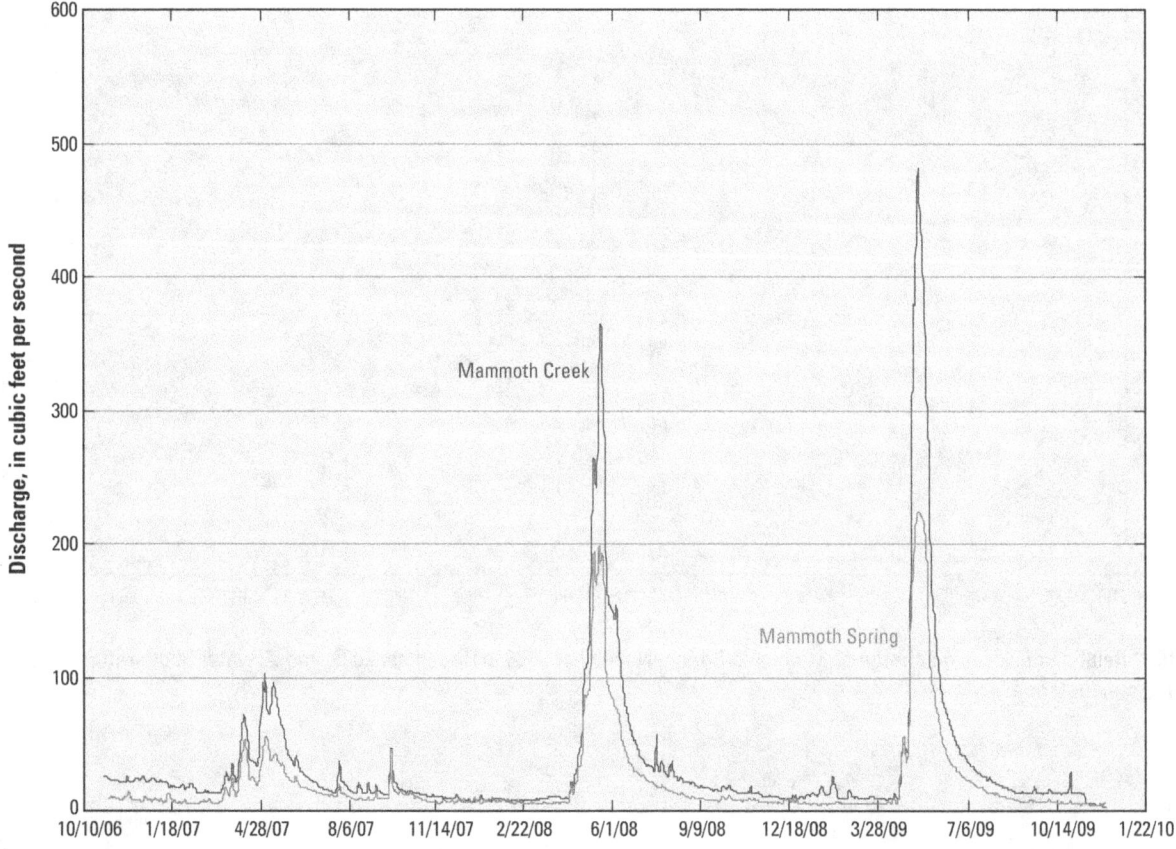

Figure 17. Relation between discharge of Mammoth Spring and total streamflow in Mammoth Creek, Markagunt Plateau, southwestern Utah, November 2006 to December 2009.

of karst springs are common and generally most noticeable at low flow, where they can represent a substantial, if not the entire, baseflow component of a spring (Spangler, 2001). Often, as could be the case with Mammoth Spring, the underflow component discharges more water than the overflow spring during baseflow conditions but does not substantially increase in flow under higher flow regimes, when the overflow spring increases to peak flow. During the snowmelt runoff period, this underflow component is obscured by the additional flow from Mammoth Creek as well as the spring.

Vulnerability of Mammoth Spring to Surface Activities

Results of this investigation show that understanding the relation between groundwater and surface water on the Markagunt Plateau is vital for effective resource management, particularly in the recharge areas for large springs. Although Mammoth Spring is situated within the watershed of Mammoth Creek, results of dye-tracer tests indicate that the greatest effects on the water quality of the spring could originate from outside the watershed boundary to the southwest of the spring at higher altitudes on the Markagunt Plateau. Activities along highways that bisect this area, particularly state highway 14, which traverses the southern part of the plateau, and highway 148 along the western margin of the plateau through Cedar Breaks National Monument, can have potentially significant effects on the water quality of the spring. These effects, which can include runoff of road salts used for deicing and spills from vehicular accidents, can be particularly detrimental where direct inputs to the aquifer exist, such as along Midway Creek, where results of dye-tracer tests and discharge-precipitation relations indicate groundwater travel times to Mammoth Spring can be thousands of feet per day during periods of snowmelt runoff and localized rainfall. Other land-use activities within the recharge area of Mammoth Spring, particularly in the southwestern part of the plateau, where numerous sinkholes serve as direct points of recharge to the underlying aquifer, have the potential to adversely affect the spring as well, which can subsequently have an impact on the surface-water drainage basin into which the spring discharges. These activities include logging, quarrying, off-road vehicle use, roadwork and maintenance, herbicide/pesticide applications, campground development, as well as encroaching residential development. Septic effluent from the latter two activities can pose an additional source of contamination. Any activities resulting in land disturbance and erosion, which can mobilize sediment, can further affect the water quality of the spring.

Although Mammoth Spring is capable of very high flows during the snowmelt runoff period, the low flows measured at the gage during baseflow conditions indicate the very low storage capacity of the aquifer supplying the spring and the high susceptibility of the spring to even short-term drought conditions. The rapid transit rates documented during this study also imply that filtration or sorption of contaminants that enter the aquifer is probably minimal, which results in the potential for adverse effects on aquatic life downstream from the spring and possible exceedances of primary or secondary water-quality standards. During periods of baseflow in the late fall and winter months, adverse effects from these activities could be greater than at high flows during the snowmelt runoff period. Although groundwater velocity generally is less during periods of lower discharge, concentrations of contaminants can be considerably higher without the effects of dilution at high flow. In addition, dispersion of potential contaminants entering the aquifer at low flow could result in longer persistence times. An understanding of the response of Mammoth Spring to precipitation events, such as documented during this study, therefore, can better prepare downstream users and others who utilize water from the spring for potential effects of anthropogenic activities and climatic events.

Summary

The Markagunt Plateau, in southwestern Utah, lies at an altitude of about 9,500 ft, largely within Dixie National Forest. The plateau is capped primarily by Tertiary- and Quaternary-age volcanic rocks that overlie Paleocene- to Eocene-age limestone of the Claron Formation, which forms escarpments on the west and south sides of the plateau. In the southwestern part of the plateau, an extensive area of sinkholes has formed that resulted primarily from dissolution of the underlying limestone and subsequent subsidence and (or) collapse of the basalt, producing sinkholes as much as 1,000 ft across and 100 ft deep. Karst development in the Claron probably began after the current altitude of the plateau was attained, when high precipitation and relief were present. Since extrusion of lava flows over the surface of the plateau during the Quaternary, surface runoff has been reduced and dissolution of the underlying limestone likely has been enhanced by high infiltration rates through the basalt.

Numerous large springs discharge from the volcanic rocks and underlying limestone on the Markagunt Plateau, including Mammoth Spring, one of the largest and most variable springs in Utah, with discharge that ranges from less than 5 to more than 300 ft^3/s. Discharge (stage) of Mammoth Spring was measured on 1- and 2-hour intervals from November 2006 to December 2009. In 2007, peak discharge of Mammoth Spring was bimodal, with an initial peak daily mean discharge of about 54 ft^3/s on April 11 followed by a second peak discharge of about 56 ft^3/s on May 4–5. In addition, an increase in discharge of more than 36 ft^3/s was documented in late September 2007, in response to a 1.1-in. rainfall event. The peak daily mean discharge was 199 ft^3/s on May 19 in 2008 and was 224 ft^3/s on May 12–13 in 2009. In both years, the rise from baseflow, about 6 ft^3/s, to peak flow took place over a 4- to 5-week period. Discharge from Mammoth Spring accounted for about 54 percent of the total peak streamflow measured at a downstream USGS gage in 2007 and 2008 and about 46 percent in 2009, while composing most of the total streamflow during the remainder of the year. Further, a significant component of total streamflow measured at the gage during baseflow conditions appears to originate from a gaining reach downstream from the spring.

Results of major-ion analyses for samples collected from Mammoth and other springs on the plateau during 2006 to 2009 indicated calcium-bicarbonate type water containing dissolved-solids concentrations that ranged from 91 to 229 mg/L. Concentrations of major ions, trace elements, and nutrients did not exceed primary or secondary drinking-water standards; however, total and fecal coliform bacteria were present in water from Mammoth and other springs. Temperature and specific conductance of water from Mammoth and other springs showed substantial variance and generally were inversely related to changes in discharge during snowmelt runoff and rainfall events. Over the 3-year monitoring period, daily mean temperature and specific conductance of water from Mammoth Spring ranged from 3.4°C and 112 µS/cm during peak flow from snowmelt runoff, to 5.3°C and 203 µS/cm during baseflow conditions. Increases in specific conductance of the spring water prior to an increase in discharge at the start of snowmelt runoff in 2008–09 were likely the result of drainage of increasingly older water from storage. Variations in water quality, discharge, and turbidity indicate a significant potential for transport of contaminants from surface sources to Mammoth and other large springs in a relatively short time frame.

Results of dye-tracer tests during this study indicated that recharge to Mammoth Spring largely originates from southwest of the spring and outside of the watershed for Mammoth Creek, particularly in areas where large sinkholes and losing streams are present. This includes Midway, Sage, and Long Valleys, and the Horse Pasture, Hancock Peak, and Red Desert areas. A significant component of recharge to the spring takes place by both focused and diffuse infiltration through the basalt and into the underlying Claron Formation. Losing reaches along Mammoth Creek are also a source of rapid recharge to the spring. On the basis of results of dye-tracer tests, maximum groundwater travel time during the snowmelt runoff period from focused points of recharge as far as 9 mi away and 1,900 ft higher than the spring, was about 7 days, indicating a velocity of more than a mile per day. Response of the spring to rainfall events in the recharge area, however, indicated potential lag times of only about 1 to 2 days, which was substantiated by changes in water-quality parameters, including turbidity. Samples collected from Mammoth Spring during baseflow conditions and analyzed for tritium and sulfur-35 showed that groundwater in storage is relatively young, with apparent ages ranging from less than 1 year to possibly a few tens of years. Ratios of oxygen-18 and deuterium also showed that water from the spring represents a mixture of waters from different sources and altitudes. On the basis of results of dye-tracer tests and relations to adjacent basins, the recharge area for Mammoth Spring probably includes about 40 mi² within the Mammoth Creek watershed as well as at least

25 mi² outside and to the southwest of the watershed. Additional dye-tracer tests are needed to better define boundaries between the groundwater basins for Mammoth Spring and Duck Creek, Cascade, and Asay Springs.

Results of this investigation show that understanding the relation between groundwater and surface water on the Markagunt Plateau is vital for effective resource management. Although Mammoth Spring is situated within the watershed of Mammoth Creek, results of dye-tracer tests indicate that the greatest effects on the water quality of the spring could originate from outside the watershed boundary to the southwest of the spring, particularly in areas of sinkhole development. Anthropogenic activities within these areas potentially can have significant effects on the water quality of the spring. Although Mammoth Spring is capable of very high flows during the snowmelt runoff period, the low flows measured at a gage downstream from the spring during baseflow conditions indicate the very low storage capacity of the aquifer supplying the spring and the high susceptibility of the spring to even short-term drought conditions. The rapid transit rates documented during this study imply that filtration or sorption of contaminants that enter the aquifer is probably minimal. An understanding of the response of Mammoth Spring to precipitation events such as documented during this study can, therefore, better prepare downstream users and others who utilize water from the spring for potential effects from anthropogenic activities and climatic events.

References Cited

Alexander, E.C., and Quinlan, J.F., 1992, Practical tracing of groundwater with emphasis on karst terranes: Geological Society of America annual meeting short course notes, Cincinnati, Ohio, October 24, 1992, 40 p.

Biek, R.F., Maldonado, F., Moore, D.W., Anderson, J.J., Rowley, P.D., Williams, V.S., Nealey, L.D., and Sable, E.G., 2011, Interim geologic map of the west part of the Panguitch 30' x 60' quadrangle, Garfield, Iron, and Kane Counties, Utah - Year 2 Progress Report: Utah Geological Survey Open-File Report 577, 1 pl.

Biek, R.F., Moore, D.W., Anderson, J.J., Rowley, P.D., Nealey, L.D., Sable, E.G., and Matyjasik, B., 2009, Interim geologic map of the south-central part of the Panguitch 30' x 60' quadrangle, Garfield, Iron, and Kane Counties, Utah - Year 1 Progress Report: Utah Geological Survey Open-File Report 553, 1 pl.

Biek, R.F., Moore, D.W., and Nealey, L.D., 2007, Interim geologic map of the Henrie Knolls quadrangle, Garfield, Iron, and Kane Counties, Utah: Utah Geological Survey Open-File Report 502.

Clark, I.D., and Fritz, P., 1997, Environmental isotopes in hydrogeology: CRC Press, LLC, Boca Raton, Florida, 328 p.

Durfor, C.N., and Becker, Edith, 1964, Public water supplies of the 100 largest cities in the United States, 1962: U.S. Geological Survey Water-Supply Paper 1812, 364 p.

Field, M.S., 2002, A lexicon of cave and karst terminology with special reference to environmental karst hydrology: National Center for Environmental Assessment, Office of Research and Development, U.S. Environmental Protection Agency, Washington D.C., EPA/600/R-02/003, 214 p.

Fishman, M.J., and Friedman, L.C., eds., 1989, Methods for determination of inorganic substances in water and fluvial sediments: U.S. Geological Survey Techniques of Water-Resources Investigations, book 5, chap. A1, 545 p.

Gregory, H.E., 1950, Geology of eastern Iron County, Utah: Utah Geological and Mineral Survey Bulletin 37, 150 p.

Heilweil, V.M., Solomon, D.K., and Gardner, P.M., 2006, Borehole environmental tracers for evaluating net infiltration and recharge through desert bedrock: Vadose Zone Journal, v. 5, no. 1, p. 98–120.

Kennedy, E.J., 1983, Computation of continuous records of streamflow: U.S. Geological Survey Techniques of Water-Resources Investigations, book 3, chap. A13, 53 p.

Meinzer, O.E., 1927, Large springs in the United States: U.S. Geological Survey Water-Supply Paper 557, 94 p.

Moore, D.W., Nealey, L.D., Rowley, P.D., Hatfield, S.C., Maxwell, D.J., and Mitchell, E., 2004, Geologic map of the Navajo Lake 7.5' quadrangle, Kane and Iron Counties, Utah: Utah Geological Survey Map 199.

Mull, D.S., Smoot, J.L., and Liebermann, 1988, Dye tracing techniques used to determine ground-water flow in a carbonate aquifer system near Elizabethtown, Kentucky: U.S. Geological Survey Water-Resources Investigations Report 87–4174, 95 p.

Mundorff, J.C., 1971, Nonthermal springs of Utah: Utah Geological and Mineralogical Survey, Water-Resources Bulletin 16, 70 p.

Spangler, L.E., 2001, Delineation of recharge areas for karst springs in Logan Canyon, Bear River Range, northern Utah, in Kuniansky, E.L., ed., U.S. Geological Survey Karst Interest Group Proceedings, St. Petersburg, Florida, February 13–16, 2001: U.S. Geological Survey Water-Resources Investigations Report 01–4011, p. 186–193. Available online at *http://water.usgs.gov/ogw/karst/kigconference/proceedings.htm.*

Spangler, L.E., Tong, M., and Johnson, W., 2005, A multi-tracer approach for evaluating the transport of whirling disease to Mammoth Creek fish hatchery springs, southwestern Utah, in Kuniansky, E.L., ed., U.S. Geological Survey Karst Interest Group Proceedings, Rapid City, South Dakota, September 12–15, 2005: U.S. Geological Survey Scientific Investigations Report 2005–5160, p. 116–121. Available online at *http://pubs.usgs.gov/sir/2005/5160/.*

Stokes, W.L., 1988, Geology of Utah: Utah Museum of Natural History and Utah Geological and Mineral Survey Occasional Paper No. 6, 280 p.

U.S. Environmental Protection Agency, 2009, Drinking water contaminants. Available online at *http://www.epa.gov/safewater/contaminants/index.html.*

U.S. Geological Survey, variously dated, National field manual for the collection of water-quality data: U.S. Geological Survey Techniques of Water-Resources Investigations, book 9, chaps. A1–A9. Available online at *http://pubs.water.usgs.gov/twri9A.*

Wilson, M.T., and Thomas, H.E., 1964, Hydrology and hydrogeology of Navajo Lake, Kane County, Utah: U.S. Geological Survey Professional Paper 417-C, 26 p., 3 pl.

Worthington, S.R.H., 2007, Ground-water residence times in unconfined carbonate aquifers: Journal of Cave and Karst Studies, v. 69, no. 1, p. 94–102.

Appendix 1.

Appendix 1.   Miscellaneous measurements of specific conductance, temperature, pH, and discharge for selected groundwater and surface-water sites on the Markagunt Plateau, southwestern Utah.

[mm/dd/yyyy, month/day/year; µS/cm, microsiemens per centimeter; °C, degrees Celsius; ft³/s, cubic feet per second; SC, specific conductance; E, estimated; ppm, parts per million; —, not measured; NTU, nephelometric turbidity units; Q, discharge; gal/min, gallons per minute; >, greater than; ≥, greater than or equal to; <, less than; ≤, less than or equal to]

Site name	Map ID Refer to Plate 1	Date (mm/dd/yyyy)	Specific conductance (µS/cm at 25°C)	Temperature (°C)	pH (units)	Discharge in ft³/s, except where indicated	Comments
Mammoth Creek springs	3	3/26/2007	188	2.9	—	—	SC 192 at 3 1°C downstream from Mammoth Creek springs
		4/19/2007	208	2.9	—	E 0 5	
		4/20/2007	210	2.7	—	—	
		6/22/2007	197	9.2	—	—	
		8/16/2007	216	10.1	—	E 1–2	Appears to be turbid and greater than flow in creek upstream
		9/13/2007	283	8.9	7.8	E 100 gal/min	No flow directly from springs; flow in channel below springs >> flow in creek upstream
		7/8/2008	205	9.4	7.7	—	
		8/2/2008	221	10.6	—	—	
		9/18/2008	212	6.9	—	—	Q of springs greater than Q of creek
		4/29/2009	178	3.6	—	—	
		6/26/2009	215	8.9	7.7	—	
		8/20/2009	230	8.8	7.7	E 0 5–0.75	
		9/11/2009	—	—		E 100 gal/min	
		4/28/2010	229	3.3	—	E ≥ 3	Flowing strong from streambed
Ephemeral spring	4	3/26/2007	189	2.9	—	E 3	
		4/18/2007	—	—	—	0.465	Q measured in morning
		4/19/2007	224	3	7 9	0.4	Q measured in evening; no flow in morning
		4/20/2007	—	—	—	—	No flow in morning
		5/8/2007	202	3.6	—	E 200 gal/min	Q measured in evening; no flow in morning
		6/21/2007	—	—	—	—	No flow in evening
		6/22/2007	—	—	—	—	No flow in evening
		8/16/2007	—	—	—	—	No flow in evening, but flowed recently
		4/18/2008	—	—	—	—	No flow in afternoon
		5/3/2008	198	2.7	—	1.28	Q measured in evening
		6/12/2008	243	6.7	7.6	62 gal/min	Q measured by R. Swenson
		7/8/2008	231	8.6	7.7	< 100 gal/min	
		7/10/2008	—	—	—	E 25 gal/min	
		8/2/2008	290	9.6	—	—	Observed intermittent flow in morning
		9/18/2008	—	—	—	0	
		4/29/2009	182	3.6	—	E 2	
		4/30/2009	—	—	—	2.65	Very murky
		5/21/2009	168	6.9	7.6	E 1–1.5	
		6/26/2009	215	8.8	7.8	E 1 5	
		8/20/2009	—	—	—	0	
		4/28/2010	266	3.7	—	E 20–30 gal/min	Time 1545; increased in flow while at site
			235	3.5	—	E 100 gal/min	Time 1640; rising on pool apparent
			225	3.5	—	E 250 gal/min	Time 1725
		5/24/2010	182	3.2	—	0.93	Appears to have been out of banks
		10/3/2010	—	—	—	0	No flow

Appendix 1.   Miscellaneous measurements of specific conductance, temperature, pH, and discharge for selected groundwater and surface-water sites on the Markagunt Plateau, southwestern Utah.—Continued

[mm/dd/yyyy, month/day/year; µS/cm, microsiemens per centimeter; °C, degrees Celsius; ft³/s, cubic feet per second; SC, specific conductance; E, estimated; ppm, parts per million; —, not measured; NTU, nephelometric turbidity units; Q, discharge; gal/min, gallons per minute; >, greater than; ≥, greater than or equal to; <, less than; ≤, less than or equal to]

Site name	Map ID Refer to Plate 1	Date (mm/dd/yyyy)	Specific conductance (µS/cm at 25°C)	Temperature (°C)	pH (units)	Discharge in ft³/s, except where indicated	Comments
Mammoth Creek above Ephemeral spring	5	3/26/2007	260	4.5	—	E 1–1.5	In 30-minute period at 1815, SC and temperature decreased to 205 and 3.7°C
		4/19/2007	198	2.6	—	—	
		5/8/2007	163	9.5	—	—	SC 169 at 2.2°C downstream from Mammoth Creek springs
		6/22/2007	198	15.1	—	—	
		8/16/2007	250	12	—	—	
		9/13/2007	—	—	—	< 5 gal/min	
		7/8/2008	208	13.8	7.8	—	
		8/2/2008	287	12.6	—	—	
		9/18/2008	413	8.2	—	—	
		4/29/2009	350	5.2	—	E 1.5	
		5/21/2009	142	5.9	8.2	—	
		6/26/2009	224	9.3	8.4	—	
		8/20/2009	—	—	—	E 100–150 gal/min	
		4/28/2010	338	5.2	—	E 100–150 gal/min	
		5/24/2010	125	4.7	—	—	
Mammoth Creek rise pool	6	11/8/2006	238	3.8	—	1.13	
		11/9/2006	239	3.9	7.8	—	
		3/26/2007	236	2.4	7.9	1.99	
		4/18/2007	—	—	—	1.37	
		4/20/2007	239	2.9	8	E 1–1.25	
		5/8/2007	235	3.7	—	E 1–1.5	
		5/8/2007	233	3.5	—	E 2	Q measured in evening
		6/21/2007	—	—	—	0.75	Q measured in evening
		6/22/2007	221	9.3	—	—	
		8/16/2007	248	9.7	—	—	Spring muddy in evening but clear earlier
		8/17/2007	233	9.7	7.6	E 1–1.5	Spring still turbid
		8/21/2007	—	—	—	0.49	
		9/13/2007	252	8.6	7.6	0.355	
		10/29/2007	249	4.7	7.7	E 50–75 gal/min	
		4/18/2008	307	2.6	7.4	≥ 1	
		5/3/2008	227	2.6	—	1.97	
		6/12/2008	206	7	7.7	1	Q measured by R. Swenson
		7/8/2008	221	8.7	7.8	0.7	
		8/2/2008	232	10.5	—	—	
		9/18/2008	221	7.2	7.7	E 200 gal/min	
		9/19/2008	—	—	—	0.51	
		10/30/2008	242	4.4	7.7	E 100–150 gal/min	
		4/29/2009	226	3	7.3	—	Strong boil
		4/30/2009	—	—	—	2.3	Very muddy
		5/21/2009	191	6.9	7.9	E 1.5–2	Turbidity 7.12, 6.90, 6 98 NTU
		6/26/2009	228	8.6	7.6	0.965	

Appendix 1.   Miscellaneous measurements of specific conductance, temperature, pH, and discharge for selected groundwater and surface-water sites on the Markagunt Plateau, southwestern Utah.—Continued

[mm/dd/yyyy, month/day/year; μS/cm, microsiemens per centimeter; °C, degrees Celsius; ft³/s, cubic feet per second; SC, specific conductance; E, estimated; ppm, parts per million; —, not measured; NTU, nephelometric turbidity units; Q, discharge; gal/min, gallons per minute; >, greater than; ≥, greater than or equal to; <, less than; ≤, less than or equal to]

Site name	Map ID Refer to Plate 1	Date (mm/dd/yyyy)	Specific conductance (μS/cm at 25°C)	Temperature (°C)	pH (units)	Discharge in ft³/s, except where indicated	Comments
Mammoth Creek rise pool—Continued		8/20/2009	236	9	7.6	183 gal/min	
		9/11/2009	224	8	—	0.25	
		11/12/2009	—	—	—	0	No flow from spring and no standing water in pit
		4/28/2010	265	3.3	—	E 1 5	
		5/24/2010	218	3.6	—	2.25	
		9/12/2010	220	7.5	—	E 1	
		10/3/2010	221	7.5	—	0.51	
Mammoth Creek above Mammoth Creek rise pool	7	11/9/2006	236	3.4	8 5	—	
		3/26/2007	362	4.5	—	< 10 gal/min	
		6/22/2007	209	9.8	—	—	
		8/16/2007	248	11.4	—	—	Muddy
		8/17/2007	—	—	8 2	—	Turbid
		9/13/2007	367	12.7	8	E 200 gal/min	
		10/29/2007	406	4.6	—	E 100 gal/min	
		4/18/2008	382	2.9	7.8	≤ 1	
		8/2/2008	248	11.7	—	—	
		4/28/2010	262	2.2	—	E 3–4	Muddy
		10/3/2010	222	8.6			
Mammoth Creek below Mammoth Creek rise pool	8	11/8/2006	—	—	—	2.95	
		3/26/2007	220	2.8	—	—	
		4/18/2007	—	—	—	8.78	
		6/22/2007	—	—	—	2.75	
		9/14/2007	—	—	—	232 gal/min	
		4/30/2009	219	5.2	—	—	
Mammoth Spring	10	9/21/2006	168–172	4.8	7.7	E 10	SC measured at several locations at springhead; dissolved oxygen 8.5 ppm
		11/6/2006	—	—	—	8.66	Measured between troll and confluence
		11/7/2006	164	4.3	7 9	—	Measured at springhead
		11/7/2006	160	4.3	—	—	Measured at troll
		11/8/2006	—	—	—	7.08	Measured 5 feet downstream of 11/6/2006 measurement
		11/9/2006	—	—	—	12.4	Measured below confluence with Mammoth Creek
		2/20/2007	159	4.3	8 3	5.8	Measured at troll; −3°C air temperature; Q measured between troll and confluence
		3/26/2007	150	4	7 9	20.7	Measured at springhead; Q measured between troll and confluence
		3/26/2007	147	4.1	8	22.4	Measured at troll; Q check measurement
		4/18/2007	—	—	—	20.7	Measured between troll and confluence
		4/20/2007	162	4.1	8 2	—	Measured at springhead

Appendix 1.   Miscellaneous measurements of specific conductance, temperature, pH, and discharge for selected groundwater and surface-water sites on the Markagunt Plateau, southwestern Utah.—Continued

[mm/dd/yyyy, month/day/year; µS/cm, microsiemens per centimeter; °C, degrees Celsius; ft³/s, cubic feet per second; SC, specific conductance; E, estimated; ppm, parts per million; —, not measured; NTU, nephelometric turbidity units; Q, discharge; gal/min, gallons per minute; >, greater than; ≥, greater than or equal to; <, less than; ≤, less than or equal to]

Site name	Map ID Refer to Plate 1	Date (mm/dd/yyyy)	Specific conductance (µS/cm at 25°C)	Temperature (°C)	pH (units)	Discharge in ft³/s, except where indicated	Comments
Mammoth Spring—Continued		4/20/2007	160	4.2	—	—	Measured at troll
		5/1/2007	—	—	—	44 2	Measured between troll and confluence
		5/8/2007	154	4.3	—	—	Measured at springhead
		5/8/2007	153	4.3	—	—	Measured at troll
		5/10/2007	—	—	—	35 9	Measured between troll and confluence
		6/20/2007	—	—	—	15.7	Measured between troll and confluence
		6/22/2007	160	5.3	—	—	Measured at springhead
		6/22/2007	157	5.1	—	—	Measured at troll
		7/30/2007	—	—	—	11.7	Measured between troll and confluence
		8/17/2007	176	5.9	7.7	—	Measured at springhead; very murky
		8/17/2007	168	5.6	—	—	Measured at troll
		9/13/2007	177	5.5	7.7	—	Measured at springhead
		9/13/2007	168	5.1	—	—	Measured at troll
		9/22–23/2007	—	—	—	—	Chris Butler (Forest Service) reports large rain event that causes spring to become turbid
		10/4/2007	—	—	—	11.8	Measured between troll and confluence
		10/29/2007	171	4.5	7.8	—	Measured at springhead
		10/29/2007	165	4.4	—	—	Measured at troll
		11/8/2007	—	—	—	6.7	Measured between troll and confluence
		2/19/2008	167	4.1	8	—	Measured at troll
		4/18/2008	180	4.2	—	E 8–10	Measured at troll
		4/18/2008	187	4.1	7.5	—	Measured at springhead
		5/1/2008	—	—	—	74 3	Measured downstream of confluence and subtracted Mammoth Creek flow
		5/3/2008	150	3.9	8.1	—	Measured at springhead
		5/12/2008	—	—	—	186	Measured downstream of confluence and subtracted Mammoth Creek flow
		5/29/2008	—	—	—	95	Measured downstream of confluence and subtracted Mammoth Creek flow
		6/12/2008	152	4.3	8.1	63 9	Measured at springhead; Q measured between troll and confluence
		7/8/2008	162	4.8	7.7	—	Measured at springhead
		7/8/2008	159	4.6	—	—	Measured at troll
		7/10/2008	158	4.7	—	—	Measured at troll
		7/14/2008	—	—	—	20 3	Measured between troll and confluence
		8/2/2008	165	5	—	—	Measured at springhead
		8/2/2008	163	4.9	—	—	Measured at troll
		9/2/2008	—	—	—	10.4	Measured between troll and confluence
		9/19/2008	162	5	8	—	Measured at springhead
		9/19/2008	159	5	—	—	Measured at troll
		10/6/2008	—	—	—	10 3	Measured between troll and confluence
		10/30/2008	167	4.5	—	—	Measured at troll
		10/30/2008	175	4.6	7.8	—	Measured at springhead
		11/18/2008	—	—	—	7.01	Measured between troll and confluence
		1/19/2009	168	4.1	—	E 4–5	Measured at troll

Appendix 1.   Miscellaneous measurements of specific conductance, temperature, pH, and discharge for selected groundwater and surface-water sites on the Markagunt Plateau, southwestern Utah.—Continued

[mm/dd/yyyy, month/day/year; µS/cm, microsiemens per centimeter; °C, degrees Celsius; ft³/s, cubic feet per second; SC, specific conductance; E, estimated; ppm, parts per million; —, not measured; NTU, nephelometric turbidity units; Q, discharge; gal/min, gallons per minute; >, greater than; ≥, greater than or equal to; <, less than; ≤, less than or equal to]

Site name	Map ID Refer to Plate 1	Date (mm/dd/yyyy)	Specific conductance (µS/cm at 25°C)	Temperature (°C)	pH (units)	Discharge in ft³/s, except where indicated	Comments
Mammoth Spring— Continued		2/4/2009	—	—	—	5.91	Measured between troll and confluence
		4/29/2009	148	3.9	7.2	—	Measured at troll; pH reading probably low
		5/4/2009	—	—	—	148	Measured downstream of confluence and subtracted Mammoth Creek flow
		5/21/2009	137	4.0–4 1	8.1	—	Took multiple SC measurements at springhead; turbidity 7.80, 7.00, 7.13 NTU
		6/11/2009	—	—	—	48	Measured downstream of confluence and subtracted Mammoth Creek flow
		6/26/2009	151	4.5	7.6	—	Measured at troll; pH reading probably low
		8/10/2009	—	—	—	12.4	Measured between troll and confluence
		8/20/2009	157	5.4	7.9	—	Measured at springhead
		8/20/2009	159	4.9	—	—	Measured at troll
		9/11/2009	165	5	—	—	Measured at troll
		9/15/2009	—	—	—	7.78	Measured between troll and confluence
		11/12/2009	164	4.4	—	5.25	Measured at troll; Q measured between troll and confluence
		12/11/2009	154	4.3	8	E 4 5	Measured at troll; Q measured downstream below confluence
		4/27/2010	153	4.1	—	22.7	Measured at troll; Q measured between troll and confluence
		5/24/2010	115	3.5	—	—	Measured at springhead
		6/6/2010	121	3.9	—	—	Measured at springhead
		9/12/2010	161	5.1	—	—	Measured at springhead
		10/2/2010	—	—	—	9.9	Measured between troll and confluence
		10/3/2010	161	4.8	7 1	—	Measured at troll
Mammoth Creek at Mammoth Spring	11	9/21/2006	258	4.3	—	E 2	
		11/7/2006	236	3.7	8 5	1.49	
		2/20/2007	—	—	—	0	
		3/26/2007	216	5.3	8 2	3.12	
		4/18/2007	—	—	—	7.64	
		5/1/2007	—	—	—	39.2	
		5/10/2007	—	—	—	31.3	
		6/20/2007	—	—	—	1.98	
		7/30/2007	—	—	—	0.45	
		9/14/2007	—	—	—	19 gal/min	
		10/4/2007	—	—	—	0.5	
		10/29/2007	—	—	—	0	Sinking upstream at campground
		11/8/2007	—	—	—	0	
		4/18/2008	—	—	—	0	E 1.5–2.0 ft³/s sinking upstream at campground
		5/1/2008	—	—	—	19.3	
		5/12/2008	—	—	—	57.5	
		5/29/2008	—	—	—	34	
		6/12/2008	178	5.2	8 2	20.7	
		7/8/2008	218	13.6	7.7	—	

Appendix 1.   Miscellaneous measurements of specific conductance, temperature, pH, and discharge for selected groundwater and surface-water sites on the Markagunt Plateau, southwestern Utah.—Continued

[mm/dd/yyyy, month/day/year; µS/cm, microsiemens per centimeter; °C, degrees Celsius; ft³/s, cubic feet per second; SC, specific conductance; E, estimated; ppm, parts per million; —, not measured; NTU, nephelometric turbidity units; Q, discharge; gal/min, gallons per minute; >, greater than; ≥, greater than or equal to; <, less than; ≤, less than or equal to]

Site name	Map ID Refer to Plate 1	Date (mm/dd/yyyy)	Specific conductance (µS/cm at 25°C)	Temperature (°C)	pH (units)	Discharge in ft³/s, except where indicated	Comments
Mammoth Creek at Mammoth Spring— Continued		7/14/2008	—	—	—	5.12	
		8/2/2008	240	13	8.6	—	
		9/2/2008	—	—	—	0.79	
		9/19/2008	229	7.1	8.6	—	
		10/6/2008	—	—	—	1.01	
		10/30/2008	—	—	—	< 5 gal/min	Main flow sinking above campground at E 200 gal/min
		11/18/2008	—	—	—	0	
		2/4/2009	—	—	—	0	
		4/30/2009	222	6	—	—	
		5/4/2009	—	—	—	59	
		6/11/2009	—	—	—	12 5	
		8/10/2009	—	—	—	< 1	
		9/11/2009	—	—	—	0	Sinking a few 100 yards upstream at E 300 gal/min
		4/27/2010	—	—	—	3.03	
		10/2/2010	—	—	—	1	
Ashdown Creek below confluence	13	6/9/2008	—	—	—	30 9	Measured by R. Riding (Forest Service)
		7/16/2008	—	—	—	7.78	Measured by R. Riding (Forest Service)
		8/14/2008	—	—	—	5.95	Measured by R. Riding (Forest Service)
Ashdown Creek above confluence	14	9/22/2008	326	12.5	8.4	E 1.5–2	
		5/19/2009	—	—	—	E 15–20	
		6/25/2009	—	—	—	E 4	
		8/19/2009	—	—	—	E 2.5	
Shooting Star Creek	15	9/22/2008	—	—	—	E 0.5	
		5/19/2009	—	—	—	E 5	
		6/25/2009	—	—	—	E 1.5	
		8/19/2009	—	—	—	E 0.75	
Jensen springs	21	9/19/2008	209	5.4	7.8	236 gal/min	Q does not include flow through pipes
		10/30/2008	219	5.9	7.9	E 200 gal/min	
		4/29/2009	207	5.2	—	E 150 gal/min	
		6/26/2009	221	5	—	E 0.75	
		8/20/2009	222	5.3	7.8	183 gal/min	Q does not include flow through pipes
		11/11/2009	216	5.2	—	E 175 gal/min	
		10/3/2010	219	6.4	—	0.475	Q does not include flow through pipes

Appendix 1.   Miscellaneous measurements of specific conductance, temperature, pH, and discharge for selected groundwater and surface-water sites on the Markagunt Plateau, southwestern Utah.—Continued

[mm/dd/yyyy, month/day/year; µS/cm, microsiemens per centimeter; °C, degrees Celsius; ft³/s, cubic feet per second; SC, specific conductance; E, estimated; ppm, parts per million; —, not measured; NTU, nephelometric turbidity units; Q, discharge; gal/min, gallons per minute; >, greater than; ≥, greater than or equal to; <, less than; ≤, less than or equal to]

Site name	Map ID Refer to Plate 1	Date (mm/dd/yyyy)	Specific conductance (µS/cm at 25°C)	Temperature (°C)	pH (units)	Discharge in ft³/s, except where indicated	Comments
Mammoth Creek below Forest Service boundary	22	11/12/2009	—	—	—	14.2	Q of 5.25 ft³/s at Mammoth Spring 2,600 feet upstream
		10/2/2010	—	—	—	19	Q of 9.9 ft³/s at Mammoth Spring 2,600 feet upstream
Tommy Creek springs outflow	25	5/9/2007	170	5.4	—	E 2	
		5/10/2007	—	—	—	1.45	
		6/22/2007	176	6.2	—	1.21	
		8/17/2007	169	5.1	8	E 1–1.5	Debris on banks indicates water has been much higher recently
		8/20/2007	168	5.1	8	—	
		9/13/2007	171	6.6	8	1.22	
		10/29/2007	169	5	8.3	E 1.0	
		2/19/2008	161	4.4	—	E 0 5	
		4/18/2008	—	—	—	E 0.75–1	
		6/13/2008	176	4.8	8.1	3.51	Q includes flow from drainage
		7/8/2008	181	5.5	8.4	2.4	
		8/2/2008	183	4.9	7.8	E 1 5	
		9/19/2008	179	5.4	8.3	1.63	
		10/30/2008	184	5.5	8.3	E 0.875	
		1/19/2009	174	4.1	—	E 200 gal/min	
		4/29/2009	136	5	—	E 2–3	Q includes flow from drainage; also took measurements of individual springs
		5/18/2009	—	—	—	E 4	Q includes flow from drainage
		5/21/2009	171	5.2	8.5	—	
		6/26/2009	193	5.3	7.9	E 3	Q includes flow from drainage
		8/20/2009	194	5.1	8.3	2.12	No flow from drainage
		11/11/2009	189	4.5	—	E 1	
		12/11/2009	181	4.3	—	E 1	
		4/28/2010	153	5.4	—	3.48	A few gal/min from drainage
		5/24/2010	224	5.6	—	E 10	Flowing strong from drainage
		10/3/2010	204	5.4	—	2.49	No flow from drainage
Long Valley Creek at swallow holes	35	6/1/2009	289	12.8	8.3	≥ 2	Flowing into at least three swallow holes
		5/22/2010	—	—	—	E 3	Flowing into swallow holes under snow
Midway Creek at swallow holes	36	4/17/2007	—	—	—	E 150–200 gal/min	
		5/3/2008	178	—	7.6	E 2	Measurements from water sample; flowing into swallow holes under snow
		6/16/2008	271	19.7	7.5	E 15–20 gal/min	Has retreated upstream to sink along hillside
		4/29/2009	—	—	—	E 10–15 gal/min	
		5/18/2009	—	—	—	E 5	
		5/20/2009	230	16.5	8.2	E 3–4	Turbidity 3.26, 3.71, 3.36 NTU

Appendix 1. Miscellaneous measurements of specific conductance, temperature, pH, and discharge for selected groundwater and surface-water sites on the Markagunt Plateau, southwestern Utah.—Continued

[mm/dd/yyyy, month/day/year; μS/cm, microsiemens per centimeter; °C, degrees Celsius; ft³/s, cubic feet per second; SC, specific conductance; E, estimated; ppm, parts per million; —, not measured; NTU, nephelometric turbidity units; Q, discharge; gal/min, gallons per minute; >, greater than; ≥, greater than or equal to; <, less than; ≤, less than or equal to]

Site name	Map ID Refer to Plate 1	Date (mm/dd/yyyy)	Specific conductance (μS/cm at 25°C)	Temperature (°C)	pH (units)	Discharge in ft³/s, except where indicated	Comments
Midway Creek at swallow holes— Continued		5/26/2009	—	—	—	3.56	
		5/22/2010	—	—	—	E 4	Flowing into swallow holes under snow
		6/6/2010	165	8.6	—	19	Swallow holes taking all flow; recently flowed on surface down valley
Stream sink along Highway 148	38	5/19/2009	—	—	—	E 4–5	
		5/20/2009	—	—	—	E 3	
		5/26/2009	—	—	—	E 0.5	
		5/28/2009	—	—	—	E 100 gal/min	
		6/1/2009	—	—	—	E 50 gal/min	
		6/6/2010	—	—	—	E 2	
Stream sink near The Craters	39	5/20/2009	59	—	—	E 100 gal/min	SC measured from water sample
Duck Creek Lava Tube outflow	44	10/30/2008	244	—	7.9	E 100 gal/min	Measurements from water sample
		4/30/2009	148	6	—	3.49	
		5/27/2009	308	6.4	7.2	1.07	
		6/26/2009	307	6.9	—	E 1	
		8/21/2009	251	8.5	7.5	197 gal/min	
		9/10/2009	—	—	—	E 300 gal/min	
		11/11/2009	250	8.2	7	E 100 gal/min	
		4/28/2010	108	6.2	—	2.23	
		5/24/2010	290	5.7	—	2.65	
		9/12/2010	230	9.1	—	≤ 1	
		10/3/2010	221	9.2	—	160 gal/min	
Navajo Lake rise pool 1	46	7/9/2008	290	5.1–5.4	—	—	SC of lake water 149
		8/1/2008	290	5.2	—	—	SC of lake water 125 at 19.5°C
		4/30/2009	245–250	3.0–3.2	—	—	
Navajo Lake rise pool 2	47	6/16/2008	218	3.5	7.3	—	SC of lake water 194 at 17.3°C
		7/9/2008	221	3.6	—	—	
		8/1/2008	219	3.6	—	—	
		9/18/2008	—	—	—	0	Lake has dropped; no flow from springs
		4/30/2009	200	3.3	—	E 1	Discharging laterally from hillside
		6/25/2009	220	3.5	7.3	—	
		8/20/2009	—	—	—	0	No flow; lake has receded
Navajo Lake rise pool 3	48	4/30/2009	202	3.3	—	—	
		8/20/2009	207	3.7	8	—	

Appendix 1.   Miscellaneous measurements of specific conductance, temperature, pH, and discharge for selected groundwater and surface-water sites on the Markagunt Plateau, southwestern Utah.—Continued

[mm/dd/yyyy, month/day/year; μS/cm, microsiemens per centimeter; °C, degrees Celsius; ft³/s, cubic feet per second; SC, specific conductance; E, estimated; ppm, parts per million; —, not measured; NTU, nephelometric turbidity units; Q, discharge; gal/min, gallons per minute; >, greater than; ≥, greater than or equal to; <, less than; ≤, less than or equal to]

Site name	Map ID Refer to Plate 1	Date (mm/dd/yyyy)	Specific conductance (μS/cm at 25°C)	Temperature (°C)	pH (units)	Discharge in ft³/s, except where indicated	Comments
Navajo Sinks	49	7/9/2008	143	20.7	—	—	Measurements made at outflow from lake
		9/18/2008	208	14.6	8.6	< 1	Measurements made at sinks
Navajo lakebed springs	50	5/24/2010	—	—	—	E 3–3.5	Total flow from six springs
Duck Creek Spring	53	5/2/2008	—	—	—	E 6–8	
		6/12/2008	270	7.9	7 9	—	
		7/10/2008	252	10.2	—	—	
		8/2/2008	230	11.7	—	—	
		9/22/2008	239	7.5	7.8	—	
		5/27/2009	254	6.3	7 9	—	
		6/26/2009	244	8.3	—	E 10–12	
		8/21/2009	214	11.1	7.8	—	
		11/11/2009	247	7.1	7.4	E 3–4	
		5/24/2010	282	5	—	—	
		10/3/2010	188	10.6	—	—	
Duck Creek Spring outflow (gaging)	55	6/9/2008	—	—	—	21	Measured by R. Riding (Forest Service)
		7/16/2008	—	—	—	12	Measured by R. Riding (Forest Service)
		9/10/2009	—	—	—	10.4	
		6/6/2010	—	—	—	32.2	Some inflow through meander bend below measurement site
Cascade Spring	56	6/12/2008	276	13	8 1	13.2	Q measured by R. Swenson; includes about 1 ft³/s from North Fork; measurements made at old USGS gaging site
		7/16/2008	—	—	—	7.98	Measured by R. Riding (Forest Service)
		8/1/2008	221	14.2	—	E 10	
		9/18/2008	256	8.7	8	E 1–2	
		5/18/2009	—	—	—	E 10–15	
		6/25/2009	245	10.4	7 2	E 7 5	
		8/21/2009	209	13.5	8.4	—	Measured at outfall (site 57); still flowing from cave
		11/11/2009	272	6.7	—	E 100–150 gal/min	No flow from cave; discharging from talus slope near confluence with tributary drainage
		6/6/2010	264	6.6	—	—	
Navajo Lake Spring	58	6/11/2008	—	—	—	E 0.75–1	
		7/9/2008	345	6.9	—	E 1	
		8/1/2008	346	6.2	—	—	
		9/17/2008	—	—	—	E 0 5	
		5/20/2009	—	—	—	E 100 gal/min	
		6/25/2009	349	6.1	—	0.72	
		8/21/2009	345	6.3	7.6	320 gal/min	
		11/11/2009	—	—	—	0	No flow; only standing water

Appendix 1.   Miscellaneous measurements of specific conductance, temperature, pH, and discharge for selected groundwater and surface-water sites on the Markagunt Plateau, southwestern Utah.—Continued

[mm/dd/yyyy, month/day/year; µS/cm, microsiemens per centimeter; °C, degrees Celsius; ft³/s, cubic feet per second; SC, specific conductance; E, estimated; ppm, parts per million; —, not measured; NTU, nephelometric turbidity units; Q, discharge; gal/min, gallons per minute; >, greater than; ≥, greater than or equal to; <, less than; ≤, less than or equal to]

Site name	Map ID Refer to Plate 1	Date (mm/dd/yyyy)	Specific conductance (µS/cm at 25°C)	Temperature (°C)	pH (units)	Discharge in ft³/s, except where indicated	Comments
Deep Creek at Taylor Ranch	59	7/31/2008	—	—	—	E 1.5	
		9/17/2008	—	—	—	E 0.75	
		5/19/2009	—	—	—	E 3–3.5	
		6/25/2009	—	—	—	E 3	
		8/21/2009	—	—	—	E 1	
Three Creeks at Larson Ranch	60	6/11/2008	—	—	—	E 2	
		7/31/2008	—	—	—	E 1	
		9/17/2008	—	—	—	E 1	
		5/19/2009	—	—	—	E 3–3.5	
		6/25/2009	—	—	—	E 2	
		8/21/2009	—	—	—	E 1